A Voyage of Discovery, into the South Sea and Beering's Straits, for the Purpose of Exploring a North-East Passage

Undertaken in the Years 1815–1818

VOLUME 3

OTTO VON KOTZEBUE
TRANSLATED BY H.E. LLOYD

CAMBRIDGE
UNIVERSITY PRESS

CAMBRIDGE UNIVERSITY PRESS

Cambridge, New York, Melbourne, Madrid, Cape Town,
Singapore, São Paolo, Delhi, Mexico City

Published in the United States of America by Cambridge University Press, New York

www.cambridge.org
Information on this title: www.cambridge.org/9781108057592

This edition first published 1821
This digitally printed version 2013

ISBN 978-1-108-05759-2 Paperback

The original edition of this book contains a number of colour plates, which have been reproduced in black and white. Colour versions of these images can be found online at www.cambridge.org/9781108057592

KOTZEBUE'S

VOYAGE OF DISCOVERY.

VOL. III.

London:
Printed by A. & R. Spottiswoode,
New-Street-Square.

I. Clark sc.

Kadu.

London, Published by Longman, Hurst, Rees, Orme & Brown, 1821.

VOYAGE OF DISCOVERY,

INTO THE

SOUTH SEA

AND

BEERING'S STRAITS,

FOR THE PURPOSE OF

EXPLORING A NORTH-EAST PASSAGE,

UNDERTAKEN

IN THE YEARS 1815—1818,

AT THE EXPENSE OF HIS HIGHNESS THE CHANCELLOR OF THE EMPIRE,

COUNT ROMANZOFF,

IN THE SHIP RURICK,

UNDER THE COMMAND OF THE LIEUTENANT IN THE

RUSSIAN IMPERIAL NAVY,

OTTO VON KOTZEBUE.

ILLUSTRATED WITH NUMEROUS PLATES AND MAPS.

IN THREE VOLUMES.

VOL. III.

LONDON:

PRINTED FOR

LONGMAN, HURST, REES, ORME, AND BROWN,

PATERNOSTER-ROW.

1821.

REMARKS AND OPINIONS,

OF THE

NATURALIST OF THE EXPEDITION,

CONTINUED.

TENERIFFE.

Many learned men have visited and described Teneriffe. Alexander Von Humboldt, and Bompland, were on this island, and ascended the Pic de Teyde. Leopold Von Buch and Christian Schmidt made the whole chain of the Canary islands the scene and subject of their scientific researches, and ascended the Pic several times; the last time, shortly before our arrival, late in the season, to collect the plants growing on it, with their fruit ripe. After the short view which was allowed us, we can only refer the curious to the learned work of Bory de Saint Vincent, and Humboldt's Travels; and to the fruits of the labours of those celebrated men whom we regretted not to have met here. The learned world now expects an account of these labours from M. Von Buch alone, as Professor

Schmidt fell, in his expedition to the river Zaire, a dear victim to the sciences which he served.*

The island of Teneriffe extends from north-east to south-west. The south-western part is occupied by the Pic and its base; the north-eastern, by rugged dreary mountains. They are separated by a broad *col* or pass; on the top of which is situated the chief city, Laguna, and below it is Santa Cruz, on the south-eastern coast, leaning on the eastern mountains. The town and harbour of Oratava lie on the opposite coast, at the foot of the Pic, in the midst of beautiful vineyards and palm-gardens, which remind us of the *insulæ fortunatæ*. The way thither, from Santa Cruz, is through Laguna, (a miserable town, with seven convents,) and through the villages of Matanza and Vittoria, names which here, as in other Spanish possessions, indicate the fate of the natives on the conquest: "Victory and Massacre!"

The country round Saint Cruz is naked and desolate; only a few palms and plantains, rising above the white walls of the town, announce to the European who lands there a more southern clime. The Flora, like that of all islands, is poor. It is most nearly related, by the similarity of its species, genera, and vegetable forms, to the Flora

* Narrative of an expedition to explore the river Zaire. London, 1818. 4to.

There has already appeared, A general View of the Flora of the Canary Islands, an Essay, by L.V.Buch, in the Memoirs of the Academy. Berlin, 1819.

of the great basin, whose waters are received by the Mediterranean: a few species of plants, of luxuriant growth, give it, however, the characteristic of the torrid zone. The date-tree, the plantain, the American aloe, the common torch thistle, and, according to Humboldt, also the dragon-tree, are strangers in this soil, as they are in Spain and Sicily. But the gigantic columnar *Euphorbia Canariensis*, with several other juicy plants, the *Cacalia Kleinia*, *Euphorbia balsamifera*, &c. belong to the rocks which they occupy, and fairly announce the vegetation of Africa.

You ask now in vain, in the garden of the Marquis de Nava, at Laguna, for the bread-fruit tree, (*Artocarpus incisa*), planted there by Broussonet himself, and which was still there when Humboldt visited the island.

Dr. Eschscholtz found, among the insects which we collected, only well-known European species. Destructive swarms of locusts (*Gryllus tataricus?*) sometimes fly from the continent and ravage these islands. We were told that, in 1811, the neighbourhood of Laguna was infested by them, and we ourselves had seen at sea, two or three degrees north of Teneriffe, and four or five degrees west of the continent, the remains of such a swarm swimming round our ship. In the sequel, one of these animals flew on board our vessel, between Teneriffe and Canaria.

The people in general appeared to us extremely

poor and ugly, but at the same time of a very cheerful temper, and very inquisitive. The Spanish gravity which they maintain, though in rags, inspired us with a certain respect. At every place we came to, we were obliged to relate our history, and produce our plants and insects. In a miserable hut in Matanza, we heard people of the lowest class converse with much sense on the volcanic nature of their mountains.

Besides our domestic animals, they have here the camel, or rather the dromedary. It is used for carrying burthens, but is spared on this rocky soil. There are two inns in Teneriffe; the one at Santa Cruz is kept by a native, the other, at Oratava, by an Englishman.

During the wars with France, 3000 prisoners of war were kept in Teneriffe. Some have remained on the island, and the children, particularly, have learnt from them a little French.

BRAZIL.

On running into the channel of Saint Catharine, you fancy yourself, on the first sight, transported into the empire of still unsubdued nature. The verdant luxuriantly-wooded mountains, which rise in unbroken lines from the shores of the island and of the continent, belong to her alone, and you scarcely observe at their foot the labours of man, who is yet a stranger there. In the interior, higher mountains, some of which take the forms of cupolas or cones, and a ridge of mountains on the continent, which are said to be slightly covered with snow in the winter months, bound the prospect towards the south.

The islands of Teneriffe and St. Catharine lie in the same latitude, one in the southern, and the other in the northern hemisphere. Yet how different is the appearance of nature in the two. There the rocky soil is only partially and scantily clothed with verdure, and foreign species of plants merely intermixed with those of Europe. Here a new creation surrounds the admiring European, in whose crowded luxuriance all is surprising and gigantic.

The kind of rock which is seen in the cliffs that rise above the water in the channel, and on the shore, is every where coarse-grained granite.

The bays are bounded by inaccessible morasses, which are covered with forests of green mangrove and towering palms. An impenetrable forest, which clothes the mountains, extends, almost without interruption, over the country. The siliquose plants, with variously feathered leaves, lofty stems, and branches spread out like a fan, seem to be predominant, accompanied however with all the usual forms of trees in rich variety. The arborescent ferns, with elegant palm like forms, attain no greater height than fifteen or twenty feet, and are hidden in the thick wood. Parasite plants (*Lianes*) of every kind (and all classes and families of plants here assume this form,) make, between the ground, the trunks, and the tops, a thickly-interwoven wonderful net. Ferns, grasses, (*Cyperaceæ*, *Heliconia*,) &c. far exceeding the height of a man, luxuriate on the ground amidst fallen trees. Another vegetable world of *Orchideæ*, *Bromeliaceæ*, *Cactus*, *Piper*, ferns, &c. wave aloft among the branches; and the *Tillandsia usneoides* hangs the crowns of aged trees with silver locks.

The paths cut in this dark wilderness are soon at an end; and he who would penetrate it, finds it impossible to reach even the top of the nearest hills.

The *Aroideæ* flourish on the sloping banks of the streams which are collected in the clefts of the mountains; gigantic *Cactus* form in places singular groups; *Bromeliæ, Orchideæ,* pepper, crown the rocks; and ferns and lichens cover tracts of dry sand. The soil, abandoned by agriculture, is soon covered with thick bushes, among which beautiful species of *Melastoma* are distinguished.

The habitations of man lie under orange groves, at the foot of the mountains; and on the shore, surrounded with plantations of pisang, coffee, cotton, &c. and by inclosures, where many of our garden plants, which have been parasitically followed by various species of European weeds, are cultivated in obscurity. The melon-tree, (*Carica papaya*,) which here shoots up in a high stem, and the *Coquero**, a kind of cocoa, with fusiform stem, and small fruit, overtop them. The genuine cocoa-palm, which grows between the tropics, does not flourish here. The Brazil or Pernambuco-wood (*Cæsalpina echinata*) enriches only more northern provinces, and the *Pilifera testiculata Bress*, should probably be likewise sought for more to the north. This is the interesting, still imperfectly known plant, the spatha of which furnishes the natural caps which are represented in Seba, (1 Tab. 3. fig. *d.*,) and may be seen in many collections.

The animal kingdom does not offer less riches, less luxuriance, than the vegetable kingdom. In harmony with the character of the vegetation, the

* There is said to be in this part a palm of this kind, the trunk of which is parted, and bears a double crown. Our time did not allow us to visit this tree.

form of creepers prevails among the birds, and many species of quadrupeds are provided with twisting tails (*Cauda prehensilis*, L.)

The most common of the monkey species here is the *Callitrix Capucina*. They are often brought up tame, under the name of Macaco. Their voice resembles that of a singing-bird. Of the quadrupeds, we also saw the aguti, and an armadillo, (*Dasypus gilvipes, sex cinctus auct.*)

Among the birds, numerous parrots and toucans are the most distinguished and frequent. There is, besides, a rich variety of species and kinds. A large grouse (*Crypturus*) is very frequent. The vultures (*Cathartes*, Ill.) clear the sea-shore, and humming-birds flutter, with butterflies, around the flowers.

Of amphibia, (the turtle perhaps excepted *,) there is an abundant variety of remarkable kinds.

Among the fish, we observed a small electric ray (*Torpedo*) without spots, and whose electric power must have been very inconsiderable, as our fishermen did not perceive it; and, among the molluscæ of the sea, a large *Aplysia*, the ink of which is used for dying red.

The greatest variety, and the greatest beauty, however, prevail among the insects. Of those which we collected, several kinds are new, and not to be found among those received from Rio Janeiro.

* The Prince Maximilian of Neuwied, who had the opportunity of a long stay in Brazil, met with sea-turtles of enormous size. — *Note of Translator.*

Among others, we found the bird-spider, (*Aranea avicularia,*) the bite of which is here considered mortal. Nature does not teach man what he really has to fear.

As soon as the sun has sunk below the horizon, luminous creatures of all kinds enlighten the air * sea † and earth. ‡ Their glittering light, the barking and cries of the amphibia, of the frog kind, and the shrill notes of the grasshoppers, give to this verdant world the animation of a scene in fairy-land.

We are indebted for our first knowledge of the natural history of Brazil to Prince Maurice of Nassau, Marcgraff and Piso, whose manuscripts and original drawings are preserved in the library of Berlin. In later times, Count Hoffmannsegg, by means of

* *Elater nocticulus* and *E. phosphoreus*, with two points of constant light on the breast-plate, and several kinds of lampyris, with light on the belly, returning at equal intervals. Their numbers, however, according to the observations of Dr. Eschscholtz, seem to be magnified in Fabricius' System, where several varieties are enumerated as species. The shining or swarming of these beetles seems to depend upon circumstances, that require further examination. Sometimes the air is filled with them, and sometimes they entirely vanish.

† Larvæ and small species of *Scolopendra*.

‡ Particularly *Medusæ*, of which we took some up on the beach, but which had suffered too much to be more accurately determined. The light was particularly visible in a wreath of points round the edge of the body, and increased by being touched, or any other excitement. The hands rubbed with the mucus of the animal, retained the phosphorescence for some time.

hunters and learned correspondence * which he kept up there, has done much for the cultivation of this branch of science; and his collections, which have been, for the most part, presented to the Berlin museum, were the chief sources for the study of it. The travels of Prince Maximilian of Neuwied, and of many zealous Germans, both men of learning, and collectors, at length finish the work; and thus, by German activity and industry, this Portuguese part of the world will be conquered for the sciences, which already owe to the Germans, Count Hoffmannsegg and Professor Link, the knowledge of the Flora and Fauna of Portugal.

The government of the island of St. Catharine, contains, as we were informed, about 30,000 inhabitants; among whom, two blacks may be reckoned to one white. We found the slave-trade still carried on here; and this government alone requires, yearly, from five to seven ships full of negroes, reckoning each at a hundred, to supply the place of those who die on the plantations. The Portuguese, themselves, import them from their African possessions in Congo and Mosambique. † The price of a man in the prime of life

* We mention here with gratitude Father Francisco Agostinho Gomez in Bahia.

† The slaves from Mosambique are the smallest number. The Guinea negroes are distinguished by the smaller angle of their profile, more projecting jaw-bones, by the deeper black of

is from two to three hundred piasters. A woman is of much less value; and speedily to exhaust the whole strength of a man, and then to supply his place by a new purchase, seems to be more profitable than to bring up slaves in the planter's house. We purposely quote the plain words of a planter in the New World, which must sound strange to the ears of an European. The sight of these slaves in the mill, where they separate the rice from its husk, in wooden mortars, with heavy clubs, keeping time in their work by a particular kind of groaning, is distressing and humiliating. These services are performed in Europe, by wind, water, and steam. We saw, also, in the village of St. Michael, a water-mill, already mentioned by Krusenstern. The number of slaves is, in proportion, smaller on the more populous islands than on the continent. Their food is meat and cassava. Those living in the houses of their masters, and such as are kept in poorer families, grow up more like human beings, than those who are compelled to work like mere machines. We were, however, never witnesses of any cruel treatment of them.

The town of Nostra Senhora de Destero, the

their skin, and many, besides, by peculiar figures, which were imprinted in their childhood, in the skin of the face and body, by a sharp instrument; marks by which the different tribes are distinguished.

residence of the Governor, lies on the island itself, on the narrowest part of the channel. The anchorage for larger ships is in its northern entrance, at a distance of several miles from the town. It contains a convent for men; and of the monks not one dedicates his idle hours to any science. Dealers in butterflies are here called naturalists.

The commerce of this colony is inconsiderable. Its harbour is only visited by American ships, to take in provisions when on their way to double Cape Horn, or go on the southern whale-fishery. Its productions are sugar, rum, rice, and coffee. Tobacco, mace, cassava (*Jatropha manihot*) fruit, &c. only for home consumption: they also grow corn, but with little success. The vine does not thrive. Both the leaves and the grapes are infected with black spots, which, with us, are ascribed to the hail. The most considerable plantations are situated on the continent, at a distance of a few miles behind the mountains.

There is no trade with the Indians; wherever any of the two nations meet, they take up their arms. Every one receives land to cultivate, and settle upon gratis, without respect to his religious opinions. Several Englishmen are said to have settled in this island, where a village is called after them.

The whale-fishery belongs to the crown. The name Armaçaõ, distinguishes the royal fisheries, which carry it on, and of which there are four in this government. They fish in the winter months at the entrance of the harbour. Sometimes only open boats go out, manned with six rowers, a mate, and a harpooner, and the fish caught is dragged on shore, where it is cut in pieces. Each Armaçaõ brings in every winter about a hundred; and we have been assured, that the number might be much greater if the payment of the wages (which are now three years in arrear) were more punctual. The whale-fishery in Brazil does not belong exclusively to this government; some lying more northward have a share in it. In this ocean, the whales of the south appear to penetrate further towards the equator than those of the north. They are said to have been met with, under the twelfth degree of south latitude.

The only vehicles which are used in the colony, and by which the produce is brought here from distant provinces, are extremely inconvenient. Two solid pieces of wood, which turn with the axle, to which they are fastened, carry a piece of wood, which serves at the same time, both for shafts and carriage, and is drawn by oxen. Horses are used only for riding. The canoes, with which the channel, the main road in the colony, is navigated, are not superior. They are long and narrow, and consist only of the trunk of a tree, hollowed out without any out-rigger. Every species of tree is used for them.

On our excursions along the coast, we found

cheerfulness, cleanliness, and hospitality prevail among a people, whose means are but scanty. We were invited into the poorest huts, where the people entertained us with fruits, and offered us meat and cassava, but refused to accept any payment in return.

CHILI.

The coast of Chili afforded us the sight of a low land, as we approached it to enter the Bay de la Conception. The peninsula, which forms the exterior edge of this beautiful basin, and the mountain-chain of the coast behind it, offer to the eye an almost horizontal line, which is not interrupted by any remarkable summit, and only the two pretty hills of the Biobio rise between the mouth of the river, after which they are called, and Port San Vincent. Whales, dolphins, seals, animated the sea around us, upon which floated the *Fucus pyriferus*, and other gigantic species, which we first met with at Cape Horn. Herds of seals basked themselves in the sun on the island of Quiquirina, at the entrance of the bay; and, in the bay itself, they swam around us as in the open sea; but no sail, no vessel of any kind, indicated that man had taken possession of these seas. We observed, only on the banks, some fields and inclosures, among forests and bushes; and low, inconsiderable huts lay scattered along the beach and on the hills.

The low mountains of the coast, from which issues the Biobio, near the town of Mocha or Conception, a broad but not deep river, conceal the view of the Cordilleras de los Andes, which rise with their snows and volcanoes at a distance of at least forty leagues from the sea, behind a broad and fruitful plain, and offer to scientific research a yet unexplored field. Molina, who has seen the Cordilleras in Peru and in this kingdom, is of opinion that these summits are superior in height to those of Quito.

The mountain, at whose foot lies the town, and on its summit the fortress, is mouldered granite, which contains undecompounded masses of the same species of rock. The hills which form the peninsula, are of schistus, over which lies red and dark-coloured clay; and the low hills, against which Talcaguano reclines towards the Port of San Vincent, consist only of strata of such clay, of which several, and particularly the upper ones, are filled with the conchylia, still living in these seas (*Concholepas Peruviana*, a large *Mytilus*, &c.) in an unaltered state. The sand of the beach and of the plain is tinged grey by fragments of slate.

The celebrated stones of the Rio de las Cruzes, near Aranco, are congeries of chiastolite. Nature on this southern frontier of Chili, the Italy of the New World, has not the unlimited productive power which filled us with astonishment at Saint Catharine's; and the mere difference of latitude does not necessarily seem to produce the difference in the two Floras. The mountains separate the countries; romantic groves of myrtle and bushes overspread the hills; and a mixture of other bacciferous trees, of congenial forms, agreeably harmonize with this predominant species. The beautiful *Guevina avellana*, of the family of the *Proteaceæ*, unites with the myrtles, and species of *Lorantus* sowed by the birds, adorn the trees and bushes with their red and white bunches of flowers. The *Fuchsia coccinea* generally fills the watered mountain clefts; a few parasite plants climb in the thicker parts of the forest; a *Bromeliacea*, the remarkable *Pitrairinia coarctata*, with rampant twisting stems and stiff leaves, covers the otherwise naked and dry hills. The beautiful *Lapageria rosea* twines round the thickets, the lighter parts of which are adorned with other liliaceous plants, amaryllis, *Alotrameria*, *Sisyrynchium*, &c.

Many European species, the *Oenothereæ Calceolaria*, *Acaeneæ*, are intermingled with new kinds; and the moist pastures of the valley are adorned, as with us, by yellow ranunculus. *

* The family of the *Proteaceæ*, and the species *Arancaria*, of the family of the *Strobilaceæ*, belong to the southern hemisphere. The species which are met with in Chili, and might remind us of *Australia*, are natives. We collected the *Goadenia repens*, which, according to Brown's observations, grows in New Holland, and Chili; it may be considered as a strand plant, like those of the *Mesembrianthemum* species which we found here, and in California; which resemble those growing in New Holland, and New Zealand, and approach very nearly the *Mesembrianthemum edule* of the Cape. We must reserve our observation on the geographical diffusion of plants, till we have arranged our botanical collections.

The winter here is not without frost, and there are instances of snow having fallen in the valley. The palm of San Jago (*Cocos chilensis*, Mol.) is not met with now so far to the south. The orange and lemon ripen indeed in the inclosed gardens of Mocha, but we do not see here the beautiful high orange groves which delighted us in Brazil. We were shown, in one of these gardens, a young date-tree, which throve admirably; and near this palm grew the *Arancariă-imbricata*, the beautiful fir-tree of the Andes, which is only met with, growing wild, in the Cordilleras, where it forms entire forests, and nourishes the inhabitants with its kernels. The Chilian strawberry, during the time of our stay, had neither blossom nor fruit.

The name of the Huemul or Guemul, (*Equus bisulcus*, Mol.,) after which we eagerly inquired, was not known to any body; and even the worthy missionary, whose conversation was so instructive to us, knew nothing of this animal. We must, therefore, leave the important zoological question, which Molina has put respecting it, to be answered by more fortunate naturalists. This author, however, appears to us to deserve little authority in natural history. We did not see, in Conception, any of the camel species of the New World; they are only to be met with in the mountains, in a wild state; and they neglect, in the total stagnation of industry, to bring them up as useful animals. We saw, in fact, no wild quadrupeds.

Screaming parrots, in numerous flocks, traverse the air: humming-birds, of various kinds, flutter round the flowers: a spur-winged water-hen (*Parra Chilensis*, Mol.) fills, with loud cries, the plain which separates the bay from Port Saint Vincent: some vultures (*Chartartes*, Ill.) seek their food on the shore; and numerous ducks, and other sea-fowl cover the sea, and settle on the banks which rise above the waves near Talcaguano.

Of amphibious animals we saw only a small frog, and a little lizard, and I believe we also perceived a snake, though Molina does not mention any.

Among the *Conchylia*, we found the *Concholepas Peruviana* and *Picus psittacus* remarkable.

We collected, among other insects, the small *Scorpio Chilensis*, which makes an exception to Molina's rule, that Chili does not contain a poisonous reptile within its frontiers. *

After the preliminary labours of Feuillée and Molina, after Ruitz andPavon, after Cavanille, who has described many Chilian plants, but sometimes confounded them, there is still much to be done

* Scorpions are, in general, less dangerous than dreaded. At the Cape of Good Hope, two very large kinds are quite common, each of which is found principally in different parts. In every place the more rare, passes for the most poisonous; and the truth is, that the sting has no more dangerous consequences than that of a wasp. Our informants spoke after their own experience. Scorpions are a favourite food of monkeys.

for the natural history of this country; and many doubts have still to be cleared up. *

With respect to the manners of the inhabitants, the obliging, incomparable hospitality of the superior classes, and the state of the colony, we can only refer to the accounts of La Peyrouse and Vancouver. We found only a change in the dress of the women, as described by the former, and of which a picture is found in his atlas. Since within these eight or ten years, it has given place to our European fashions, after the newest of which the ladies eagerly inquired; and the dress of the men is only distinguished by the Araucanian poncho, and the broad-brimmed straw hat. †

But, amidst the cheerful and easy society which

* Louis Feuillée, Journal des Observations Physiques, Mathématiques, et Botaniques, faites dans l'Amérique Méridionale. Paris, 1714—25, 4to.

Molina, Saggio sulla Storia Naturale del Chili. Bologna, 1782. 8vo. Secunda Edizion Bologna, 1810. 4to. does not clear up what was left obscure in the first edition.

Ruitz et Pavon, Floræ Peruvianæ et Chilensis prodromus. Madriti, 1794. Romæ, 1799. Systema Vegetabilium Fl. Per. et Chili. Madriti, 1798.

Flora Peruviana et Chilensis. Mad. 1798 et 99. The *Eryngium rostratum*, Cav. is not the *Eryngium* growing near Talcaguano.

† The poncho, is a longish four-cornered covering, striped, length-wise, with riband, like ornaments, of a particular kind of woollen cloth, in the middle of which is a slit to put the head through; the two ends hang down before and behind. Chili formerly received the fashions from Lima; but the Chilian poncho is worn even in Peru.

we enjoyed in Conception, we could not refrain from melancholy reflections on the political crisis in which this country is engaged.

He who enters neutral between two parties in a civil war, sees only, in the crowds on both sides, wild intoxication and hatred. We saw only the royal party, which the liberals, remembering the history of the mother-country, call Moors. Compared with the numerous splendid female circles, we saw only a few men, officers and functionaries of the king, and a ragged, miserable, wretched, and motley soldiery.

Many individuals of the patriot party, which was then oppressed, were in the state prisons, which had been enlarged, by adding a church to them; and were employed in building the fort, which was erecting to keep the city in awe. Some were sent to the island of Juan Fernandez; others among them, and many of the clergy, had assembled in Buenos Ayres, under the flag of their native country, which, after the fall of Carthagena, which we saw celebrated with enthusiastic joy, was represented to us as entirely vanquished.

And Chili, which Molina describes as a terrestrial paradise, where a fruitful soil is adapted to every species of cultivation, whose riches in gold and silver, corn, delicious wine, fruits, productions of all kinds, timber, oxen, sheep, and horses, are immense, languishes in fetters, without navigation, commerce, or industry. The smuggling trade of the Americans, whose agents are the monks, supplies it with all necessaries, but only for ready money, without its turning its own productions to advantage; and these same Americans, exclusively, carry on the whale-fishery on its coasts.

History has decided on the revolution to which the United States of America owe their existence, their prosperity, their rapidly increasing population and power; and all the nations of Europe regard with undissembled favour the struggle of the Spanish colonies. The separation from the mother-country is to be foreseen; but it is doubtful when a wise, tranquil development will seal the transition from oppression to independence.

The town of Mocha is regular, and of great extent, but the houses low, and large, and provided with windows only towards the interior court-yard. They are indeed built in a manner well-adapted to frequent and violent earthquakes, but not for the cold of winter. They are not acquainted with either chimnies, or stoves. The poorer class have not even got a kitchen hearth, and dress their victuals in the open air, or under the entrance-hall. In the evening many fires burn in the streets of Talcaguano, at which the people warm themselves; and we were ourselves witnesses of a fire caused by this custom, which reduced a house to ashes. The vineyards, which produce the delicious Conception wine, lie at a considerable distance from the town. The wine is brought, like the corn, in leathern sacks, and it is preserved in large earthen vessels: there are no barrels. Beasts of burthen, asses of a very fine race, and mules, supply the place of carriages, of which there are but very few, and of the same kind as those in St. Catharine's. The Governor-Intendant alone has a chaise, made in Lima, which he seldom or never uses. The horses are very fine, and good, and riding is quite general; the women also ride in their journies, or make use of carts resembling our shepherds' huts, which are drawn by oxen.

The creole is always on horseback; the poorest possesses at least a mule, and even a boy rides behind the asses which he drives. The noose, or *lazo*, is in general use.

We will mention a custom singularly founded on religious ideas, which was offensive to our feelings. When a child dies, after having received baptism, the evening before the burial, the corpse itself is dressed up as the image of a saint, and placed erect in a lighted room, on a kind of altar, which is adorned with burning tapers, and garlands of flowers. The company then assemble, and they amuse themselves during the night, with wordly songs, and dancing. We twice witnessed such festivals in Talcaguano.

Several Araucanians whom we saw in Conception, as they belonged to the lowest class of their people, who hire themselves to the Spaniards as day-labourers, could not give us a faithful portrait of that warlike, eloquent, strong, and unmixed nation, whose feelings of liberty, and skill in the military art, opposed an insurmountable bulwark, first to the arms of the Incas, and afterwards to the desolating conqueror of the New World. The Peruvians did not penetrate farther south into Chili than to the river Rapel, and the Biobio has remained the proper boundary of the Spaniards, who, more to the south, possess only the places of St. Pedro, Aranco, Valdivia, the archipelago of Chiloe, and inconsiderable frontier-posts, the road to which leads through the independent country of the Indians.

We will not transcribe from other books on the history of Chili, and its nations, as every person is able to procure them. *Ovalle is faithful, detailed,

* Ovalle (P. Alonzo) Breve Relacion del Reyno de Chili, 1646. Molina, Saggio sulla Storia Civile del Chili, 1787-8.

The Abbé Giovanni Ignazio Molina, a native of Chili, is reckoned among the principal authors of the Italian literature. We regret that his historical work has not, as well as his natural historical work, been translated into German. In it, may be found, a Catalogo di Scrittori delle cose del Chili; an appendix to it in Mithridates, 3d part, 2d division, p. 391. and the following; and in Linguarum totius Orbis Index, J. S. Pater, Ber. 1815. p. 18.

Among the means for learning the Araucanian language, we particularise B. Havestadt Chilidugu, Monast. 1777, which, being more accessible than the various editions of Louis de Valdivia, published at Lima, may be obtained by other philologers as well as ourselves. Molina gives, in his Saggio sulla Storia Civile, a very accurate and clear picture of this beautiful and diffusive. Molina writes, with an attachment to his country, a history which cannot be read without interest; and truly the history of a people which is still in the state where man is appreciated at his just value, and appears in independent greatness and energy, must be more attractive than those civilized states where calculation presides, the character disappears, and man only weighs, or is weighed.

Among the authorities for the history of Chili, are several Spanish epic poems, of which the Araucana, by Don Alonzo de Ercilla, has the first rank. This work is mentioned with honour in Don Quixote; Voltaire has praised it; and an edition of it has appeared in Germany, (Gotha, 1806-7.) This elegantly-versified historical fragment, whose author celebrates wars in which he himself fought, deserves less the attention of the German literati than of the inquirer into history. Historians refer to it with confidence, and in Chili, where it is considered as a national poem, it is more read than any other book.

As a supplement to the historians of Chili, we communicate the notices for which we are indebted to Father Alday, a missionary, who spent a part of

---

language. We have found occasion in another place, to mention and compare the nations and languages of South America, with those of the islands of the Great Ocean, and of the eastern part of Asia; and observe, that our researches have not led us to find a community between them.

his life among these people, and add only a few more observations.

The last convention between the Spaniards and the Indians was concluded anno 1773. The latter have had, since that time, a resident with the Captain General of Chili, at San Jago, and peace has not been interrupted. La Peyrouse seems to have been purposely deceived, to prevent him, or the gentlemen of his expedition, from making an excursion into the interior of the country. They made him believe that a war was carrying on, of which history is ignorant. They told us that, under present circumstances, the Indians were faithfully attached to the King of Spain, and that they defended the defiles against those of Buenos Ayres. The direct communication of the colony with the mother-country which formerly went over the Cordilleras, by way of Mendoza, the Pampas, and Buenos Ayres, was, during our time, renewed by way of Lima and Carthagena. A parliament, a solemn popular assembly of the Indians, at which the Captain General appears in person for the Spaniards, where the interests of the two nations are discussed, and the bond of friendship confirmed, was to be held in a few weeks, at the usual place on the frontiers, Los Angeles; and we regretted extremely that we must miss this opportunity of witnessing the large assembly of a free people, whose history, though recorded by their hereditary enemies, is so distinguished for great men and noble actions.

NOTICES OF THE MISSIONARY, FATHER ALDAY.

TRANSLATED FROM THE SPANISH MANUSCRIPT.

The history of the kingdom of Chili was written from the beginning, by Garcilaso de la Vega, mixed with his history of Peru. The celebrated Ercilla commemorated it in heroic verse, to the end of his own mission. Father Ovalle, at Rome, wrote an excellent account of the deeds and misfortunes of this kingdom, from its foundation up to his time; and, lastly, the work was finished by Abate Molina, who wrote and executed this history in all its parts. This learned ex-jesuit treats, admirably, of the mineral and vegetable kingdom; so that nothing can be added to what he has said. The riches which Chili contains in its bosom are inexhaustible; its soil is the best adapted to all the productions which enrich Europe, as it enjoys an equal degree of temperature on its extreme frontiers; it has neither the storms that are fatal to the silk-worm, nor any hail to injure the fruits of the earth. No beast of prey haunts its mountains which could threaten the inhabitants, and not a single poisonous reptile is found within its limits.

The Indians who inhabit the country, from the river Biobio to Osorno, are divided into four provinces, which extend like four zones, from north to south. They are about 80,000 in number. They are generally above the middle size, strong and robust, and very active. They are extremely addicted to liquor *, and this is the principal reason of the decrease which we observe when we compare the present population with that which history records at the time of the conquest. An acute observer says, that Don Garcia Hurtado de Mendosa waged the most terrible war against them when he gave them the apple-tree. These trees now form large groves in their territory. The blood of the Indians is now no where found unmixed. This arises partly from the Spaniards, who fly to them to escape justice; partly from the women whom they made slaves, on the destruction of seven colonies, on different occasions during the war; and partly from the Dutch, who deserted, in such great numbers from the Dutch expedition, which landed at Valdivia, in the reign of Philip IV., that the commander, on his return, was obliged to sink two galleons, not being able to man them. The descendants of these Dutch are now seen in Villarica and Tolten, to the shores of Rio de la Imperial. †

* Their intoxicating drink is cyder; even the poor Creoles prepare and drink it.— *Translator.*

† The accounts which we have of the Dutch expedition to Chili, in 1643, under Hendrick Brouwer, are in direct contradiction to the facts mentioned here. Compare Burney's Chronological History, vol. 5. p. 113. Molina only slightly mentions this circumstance.

The land of the Indians, according to the latitude, is as fruitful as that of the Spaniards. But, on account of the very diminished population, we see many fields covered with high trees and low bushes, whose level ground convinces us that they once belonged to agriculture, and which show evident marks of having lost their former inhabitants.

The numerous species of trees which grow in the country of the Indians, as well on the plain as on the precipices of the Cordilleras, are also met with in the Spanish territory: only the Taÿo is an exception. The bark of this tree, which is smooth, and about the thickness of a line, is very efficacious for the cure of the internal aposteme, and every kind of ulcer or wound. They drink water, in which it has been boiled, for these diseases; and bathe and wash in this water for similar external complaints, and then strew themselves with the powder of the same bark, which is dried and rubbed. The rest of the plants and herbs of this district are of the same nature as those produced in the Spanish territory.

In the mountains, lions are met with, which prey on other animals, but do no injury to man, whom they avoid. There are also several mountain-goats and deer, of the size of a lamb: their flesh is of a good taste. The rivers abound in fine trout, and smaller species of fish. On their banks an animal is met with, though not frequently, which lives on fish, and is called by the Spaniards, water-cat, and by the Indians, *guillin*. Its skin furnishes a most valuable fur, and its extremely fine hair has not its equal for the manufacturing of hats, (*Castor Huidobrius*, Molina.)

But let us return to the Indians. They make use of a very cautious policy to preserve the independence of their states. They will suffer no Spaniard or stranger to travel through their territory, much less to explore it, without the permission and consent of the Cacique of the district; which permission he never gives, unless he knows to whom it is given. This is likewise observed with respect to the missionaries who travel in the interior of the country without being accompanied by the missionary of the district himself; for the distrust of the Indians does not go so far as to make use of these measures of precaution towards them. Most of the Indians are Christians, and all, without exception, desire, and wish that their children should be baptized; but they refuse, when they come to an age to receive Christian instruction, to present them to the church; because, say they, if the missionaries make themselves masters of the children, they would also make themselves masters of the parents; and they would, consequently, lose the political liberty of their fathers. Therefore, in the annexed tables, only those Indians will be mentioned, who live in the missions as children of the church, and not those who are mixed with the heathens of the district.

For the rest, we refer to the account of Thomas Falkener, published in London, anno 1774; this Englishman lived for forty years in Paraguay, the kingdom of Chili, and on the coast of Patagonia.

The division of the Indians into four provinces has been already mentioned. They are the Araucanians, the Llañistas, or inhabitants of the plain, the Huylliches, and the Pehuenches. The Araucanians inhabit the coast, divided into the following governments: Arauco, which gives the name to the whole province; Tucapen, out of which they have always chosen their leaders for their greatest undertakings, Lleulleu, Tixua, Imperial Baxa, Voxoa; Tolten, where the jurisdiction of Valdivia begins, Maxiguirra, Valdivia, Cuidico, Cumcos. Each government has its first Cacique, who governs all the districts comprised in his territory. Over each district presides an Indian of consideration, with the title of Guilmen. The dignities of Cacique and Guilmen are hereditary. The same division in the government, and district, and the same names of Cacique and Guilmen, are found in the three other provinces; among the Llañistas, inhabitants of the plain; the Huylliches, inhabitants of the declivity of the Cordilleras; the Pehuenches, inhabitants of the vallies, hills, and interior vallies. No Cacique or Guilem interferes with another's territory. They call together provincial assemblies to transact important business; those of the coast from Arauco to Tolten, in Chili, and those from Tolten to Cumcou, in Valdivia. The greatest harmony prevails among them. The Caciques come alone, attended by a few warriors to the provincial assemblies; but if the business concerns the whole country, deputies from the other provinces take a part in the deliberations, after the affair has been discussed in the assembly of each province. All the Indians, except the Pehuenches, cultivate the ground, and sow wheat, maize, barley, beans of different sorts, and flax, of which they eat the seeds, and use the straw for brooms. They all possess horses, oxen, sheep, hogs, and fowls; mules are very rare. They neither plant nor sow vegetables or fruit-trees. Oxen and horses alone spread the seeds of the apple-tree. The Pehuenches have many studs, which furnish them with flesh and milk for food; for though they keep oxen and sheep, they never eat their flesh. They manufacture the wool of their sheep themselves, and sell the oxen to the Spaniards. The women are in general very industrious, assist their husbands in the labours of the field, and live so much in submission to them that the penance which God laid upon the first woman is here fully developed.

TABULAR VIEW

Of the Missions of the College de Propaganda Fide of Saint Il de Fonse, of the City of Chillan, in the kingdom of Chili, and of the fruits reaped by them since they have been administered by the said College, with notices of the year of their foundation, and the numbers of the Missionaries employed in each.

DRAWN UP IN THE YEAR OF OUR LORD 1815.

MISSIONS.	Year of Foundation.	Number of the Missionaries.	BAPTISMS.		MARRIAGES		BURIALS.		Christians of every rank, age, and sex.	Heathens of every rank, age, and sex.
			Children.	Adults.	Solemnized.	Existing.	Children.	Adults.		
V. † Valdivia -	1769	3	1113	361	460	79	469	465	465	20
† Mariquina -	1769	2	1039	147	342	130	243	140	775	466
Arique - -	1776	2	1016	79	235	88	410	246	487	45
Niebla - -	1777	2	445	50	123	48	170	120	364	4
Nanihue - -	1777	2	406	68	137	58	97	105	264	20
Quinchilca - -	1778	2	1035	167	246	120	265	163	622	200
Rio Bueno - -	1778	2	991	241	239	181	186	80	1086	500
Dalli Pulli -	1787	2	1219	248	245	260	250	113	1216	303
Cudico - -	1787	2	1406	185	215	159	326	80	1231	275
O. Quilacahuin -	1794	2	730	157	106	90	102	30	667	377
Cuyunco - -	1794	2	882	272	180	171	150	77	964	890
Costa - -	1806	2	635	130	96	85	166	40	949	450
C. † Arauco -	1768	2	1016	66	201	106	282	202	540	2300
Tucapen - -	1779	2	108	12	17	5	18	8	15	6400
St. Barbara -	1758	2	80	16	10	2	26	54	8	150
		31	12121	2167	2852	1622	3160	1787	9644	12400

SHORT NOTICES

Of the Missions which have been lost, the year of their foundation, and when they were withdrawn, and of the fruits which they have reaped.

MISSIONS.	Year of their Foundation.	Year of their Loss.	Baptisms.	Marriages	Burials.
P. Culiaco - -	1758	1766	59	6	26
P. Rorinlern -	1758	1766	—	—	—
P. Lolip -	1766	1766	52	—	—
C. Imperial Buxa -	1768	1787	4	—	—
V. Tolten el Baxo -	1776	1787	179	6	6
			294	12	32

MISSIONS.	* Geographical Situation. Latitude.		Longitude.		Extent. N. S.	E. W.	Distance from the College.	District.
	°	'	°	'				
Valdivia -	39	47	302	28	6	7	160	10
Mariquino -	39	24	302	31	6	7	140	10
Arique -	39	49	302	48	4	5	155	8
Niebla - -	39	49	302	32	9	2	160	6
Nanhiue	39	32	302	48	10	8	145	9
Quinchilca -	39	42	303	18	13	10	179	12
Rio Bueno -	40	29	303	24	7	8	190	12
Dalli Pulli -	40	18	303	21	7	8	187	8
Cudico -	40	15	303	18	4	4	185	7
Quilacahuin -	40	27	303	18	6	4	193	6
Cuyunca -	40	36	303	21	8	7	199	7
Costa - -	40	37	302	47	7	4	201	6
Arauco -	37	21	302	30	20	4	50	16
Tucapen -	37	56	302	30	18	6	70	24
St. Barbara is only an hospital without cure of souls.	36	41	304	2	—	—	40	—

* Astronomically determined by Cedillon.

Remarks.

The missions marked †, owe their foundation to the Jesuits, and came into the hands of the Franciscans in the years mentioned in the table. Those marked with letter V, are under the jurisdiction of Valdivia; those marked O, under the jurisdiction of Osorno; and those with the letter C, in the jurisdiction of Chili. All are properly missions, except St. Barbara, which is an hospital for those who are destined for the conversion of the Pehuenche nation, which inhabits the Cordilleras. The Franciscans had three missions there, which are marked in the second table with the letter P, together with the years when they were lost. For want of missionaries they have not been re-established, though, in 1803, the Indians applied for them; for they well know how useful it is to them to have missionaries among them, to help them, and check their rage for blind heathenism. In the same table the missions marked C, belong to Chili, and those with V, to Valdivia.

The three first missions in the second table are situated in the Lower Alps of the Cordilleras de los Andes, where, from the source of the river Nuble, to the Archipelago of Chiloe, are the following volcanoes: Chillan, Antuco, Callagui, Chandel, Villarica, Huanchue, Copi, Llanguihue, and Purarauco. It is to be observed, that at the foot of each of these volcanoes there is a large lake, and that the principal rivers of this extensive country rise from these lakes. Thus from Chillan, or out of its lake, flows the river Nuble; from Antuco, the Laxa; from Callagiu, the Biobio; from Chandel, the Imperial; from Villa-rica, the Tolten; from Huanchue, the river of Valdivia; from Copi, the Rio bueno; from Llanguihue, the Pilmayguen; and from Purarauco, the river Rauhue, which waters the territory of Osorno, and, half way to Chiloe, forms a second branch, which is called Maypuhue.

The Indians who inhabit the Cordillera are called Pehuenches, a name which originates from the fir-trees which abound there. They are extremely robust, and inured to heat and cold beyond all conception; they are likewise very brave and bold, and are much feared by the inhabitants of the valley. Their usual diet is horse-flesh, and the kernels of pine-cones, which the forests produce in great abundance. They do not sow seeds of any kind, and when they want vegetables they exchange for them with the Indians in the plain, salt and the kernels of pine-cones; they carry on the same bartering trade with the Spaniards out of the territory of the Cordilleras. They possess extremely rich salt springs, which extend two days' journey from east to south, without a single drop of sweet water being found in this immense tract. The salt, which is very wholesome, is as white as snow, and it is very easy to reduce it to a powder as fine as flour.

The women, who are very industrious, weave a great number of ponchos, while the men finish, at times, and, as it were, for recreation, troughs and other articles of wood. This industry is the fruit of their intercourse with the Spaniards. Among the few different kinds of trees which the Cordillera produces, the fir-tree is the most esteemed. This tree grows to the height of twenty-five varas, (about seventy-five feet), and its thickness is in proportion to its height. It is probable that if the least attention were paid to it, it would surpass all other wood for ship-building. The Pehuenches extend their intercourse with the Spaniards beyond the Cordillera to Buenos Ayres. They formerly made incursions in the Pampas, plundered travellers, penetrated into the small villages and settlements of the Spaniards, murdered the men, and carried off the women and children, whom they treated as slaves. The missionaries have redeemed several of these unfortunate people from them, and set them at liberty. The Pehuenches are now kept in check by the two forts of St. Juan and St. Carlos, which the people of Mendosa have erected in proper places.

CALIFORNIA.*

A LOW ridge of mountains borders the coast of California, where we saw it, and intercepts the prospect into the interior of the country. It has not a volcanic appearance.† The harbour of San Francisco, in which Burney, (Part I. p. 354.,) with learned criticism, recognises the harbour of Sir Francis Drake, enters through a narrow passage, receives some rivers from the interior, branches out behind the eminences, and forms into a peninsula, the country lying south of the entrance. The Presidio and the Mission of San Francisco lie on this tongue of land, which, with its hills and downs,

* For accounts of California, see Noticia de la California y de su Conquista, por el P. Miguel Venegas. Madrid, 1775. 4to. of which a Natural and Civil History of California, London, 1759, (*Quære?*) is a translation.

Diario Historico de los Viages deMar y Tierra hechos al Norte de la California. D. Vicente Vila. Mexico, 1769. Accounts of the American Peninsula California, by a Priest of the Society of Jesus, who has lately resided there many years. Mannheim, 1773. And the Voyages of La Peyrouse, Vancouver, and Langsdorff.

† Near St. Barbara (34° north latitude) there arises from the coast a still burning volcano, the foot of which is washed by the sea; and in other places of the peninsula, there are proofs of a volcanic nature.

was the narrow field which lay immediately open, to our researches.

The hills on the northern side of the harbour are composed of flinty slate. The hill opposite to them on the southern side, and on which the fort lies, is of serpentine. On going along the strand, towards the south, to the Punta de los Lobos, the serpentine ceases, and you meet several almost perpendicular strata of flinty slate, which rest against coarse-grained sandstone, veined with calc-spar; and this sandstone, of which the more southern hills to the Punta de los Lobos consist, seems to be the kind of rock that lies the lowest. Quicksand lies in many places at a considerable height over the stone, and in many places new sandstone has been formed.

The environs of San Francisco, in the northern hemisphere, are much poorer in natural productions than the coast of Chili, under the same latitude, in the southern. In the spring, when winter has afforded the earth some moisture, the hills and valleys are indeed adorned with brilliant iris and other flowers; but the drought soon destroys them.

The fogs, which the prevailing sea-winds blow over the coast, dissolve in summer over a heated and parched soil, and the country exhibits in autumn only the prospect of bare scorched tracts, alternating with poor stunted bushes, and in places, with dazzling wastes of drift sand. Dark pine forests appear here and there on the ridge of the mountains, between the Punta de los Reyes and the harbour of San Francisco. The prickly-leafed oak, *Quercus agrifolia*, is the most common and largest tree. With crooked boughs and entangled branches, it lies, like the other bushes, bent towards the land; and the flattened tops, swept by the sea-wind, seem to have been clipped by the gardener's shears. The Flora of this country is poor, and is not adorned by one of those species of plants which are produced by a warmer sun. It however offers much novelty to the botanist. Well-known North American species* are found mixed with others belonging to the country†; and most of the kinds are yet undescribed. Only Archibald, Menzies, and Langsdorff, have made collections here; and the fruits of their industry are not yet made known to the world. The season was not very favourable for us. We, however, gathered the seeds of several plants, and have reason to hope that we shall be able to enrich our gardens with them.

These deserts serve for the retreat of numerous animals, of which many are probably not yet de-

* *Ceanotus, Mimulus, Oenothera, Solidago, Aster Rhamnus, Salix, Aesculus?* &c. Species of wild grapes, which we did not see ourselves, are said to be very abundant in the interior, and to produce fruit of an agreeable taste.

† *Abronia, Eschscholzia, Cham.*, and others which are to be now described.

scribed. They are called here by the names of well-known kinds; small lions, wolves, and young stags, goats, and rabbits. But their most terrible inmate is the bear, which, according to the accounts of the hunters, is uncommonly large, powerful, savage, and tenacious of life. He attacks men and animals, though there is no want of vegetable food, and collects, in countless troops, round the dead whales that are cast on the beach. The colour of his fur changes from brown into very light, and is frequently spotted with other colours. It does not appear to be the white bear of Lewis and Clerke, and is not the well-known American black bear. To judge by the one we saw, (a young female,) we could not distinguish it from the European brown bear, and the skull, which Professor Rudolphi has examined, seems to him to belong to this species. The Spaniard is very skilful in catching this dangerous beast with the noose, and takes pleasure in its combat with the bull. The whales and seals of the north visit this coast; the sea-lion is common, and the sea-otter now no where more frequent than here.

There is an uncommon number and variety of birds, the *Oriolus phœniceus* is found in innumerable flocks. We did not see a single kind of the family of creepers, and a splendid humming-bird seemed to be a stranger which had strayed hither from the south.

Melancholy feelings attend our offering a few

words on the Spanish settlements on this coast. With an avaricious thirst for possession, Spain extends her territory here, merely because she envies others the room. She maintains her Presidios at a great expence, and tries, by the prohibition of all trade, to force ready money back to its source. But a little liberty would make California the granary and market of the northern coasts of these seas, and the general resort of the ships which navigate them. Corn, oxen, salt, (at St. Quentin, Old California,) wine, the produce of which would increase the demand, give it in many respects a superiority over the Sandwich islands, though their situation, on the route between China and the north-west coast, is more advantageous. But industry and navigation, the offspring of liberty, would speedily transfer a profitable share in this trade to California, which possesses the sea-otters in greater abundance than all the other coasts.*

Yet California lies without industry, trade, and navigation, desert and unpeopled.† It has remained neglected, without any importations from Mexico, during the six or seven years of the war between Spain and its colonies. The ship from St. Blas, which formerly brought supplies to

* The Californian sea-otter skin is indeed inferior to the northern, but the difference is not very great.

† As a proof, it may be mentioned, that the cwt. of flour costs, in the missions of this country, six piasters; while at San Blas it costs forty, and at Acapulco fifty.

these settlements yearly, arrived in Monterey only while we were there. The missions possess some bad barks in the harbour of San Francisco, built by foreign captives. Even the Presidio has not a single boat; and other havens are no better off. Strangers catch otter-skins even in the Spanish harbours; and only a smuggling trade, which the new governor of California, since his appointment (fourteen months ago) has tried to suppress, furnishes this province with the most indispensable articles. Spain has given way in the affair of Nootka. England and the United States, without regarding its vain territorial possessions, are now negotiating about the colony at the mouth of the Columbia; and the Russian American Company have still a settlement a few leagues north of San Francisco.

But the maintenance of this colony is ascribed to another motive besides policy: namely, the pious intention of propagating the Christian religion, and the conversion of heathen nations. The governor of the province himself, informed us, that this was the real state of the case. Well, then, a good work has been here injudiciously begun and ill-executed.

The pious Franciscans, who hold the missions in New California, are not skilled in the arts and trades which they ought to exercise and teach, nor in any of the languages spoken by the nations to whom they are sent. They are monks, exactly like those in the convents of Europe. * They direct a considerable agricultural establishment; (always two in each mission,) perform divine service, and converse with those committed to their charge, by means of interpreters, who are themselves Indians. All property belongs to the community of the mission, and is administered by the fathers. The savage Indian derives no immediate advantage from his labours; no wages, if he happens to be let out as a day-labourer on the Presidio. The mission receives the money which he earns. He acquires no notion of property, and is not bound by it. We do not deny the mildness, the paternal anxiety of the missionaries, of which we have several times been witnesses. † The relation still remains what

* We were more offended than edified by a sermon preached in the Spanish language, in the mission of San Francisco, on the Saint's day; and in which the patron saint was placed on an equality with Christ.

† The following is one exâmple: The fathers sent the Indians in their boat to our anchoring-place, merely that they might look at our ship, which was a new object to them. The Indian, in the mission, dances his national dances, on Sunday, in presence of the fathers, and plays, always for gain, his usual game of chance; he is only forbidden to stake his coat, a piece of coarse woollen cloth, manufactured in the mission: he can also enjoy the hot-bath, to which he has been accustomed. The dances are boisterous, different in each tribe, and the tune generally without words. The game is played between two antagonists, at " odd or even," with short sticks; an umpire keeps the account with other sticks. The usual bath of the Indians, like that of most of the northern nations, is the following: at the entrance of a cave on the sea-shore, in which the bathers are, a great fire is made; they suffer it to go out,

is here represented; and, in our opinion, it would differ only in name, if the master of slaves kept them to work, and let them out at pleasure, he also would give them food.

The savage comes unthinkingly into the mission *, receives the food which is willingly offered him, and listens to the instructions: he is still free. But as soon as he is baptized, he belongs to the church; and hence he looks with pain and longing to his native mountains. The church has an inalienable right to her children, and exercises this right with rigour.

The savage is inconsiderate and inconstant, like a child. Work, to which he is unaccustomed, is too difficult for him; he repents of the step which binds him, and demands his pristine liberty. The love of home, is, in him, a ruling passion. The fathers allow their Indians, for the most part, twice a year, a leave of absence for some weeks to visit their friends, and their native place. † On occa-

---

when they have perspired sufficiently, and then leap over it, and plunge into the sea. A vapour bath, resembling the Russians was formerly used by most European nations. Erasmus Roterodamus Coll. Diversoria. Atque ante annos viginti quinque nihil receptius erat apud brabantos quam thermæ publicæ, eæ nunc, frigent ibique, scabies enim nova docuit nos abstinere.

* There is no fixed boundary to the different missions. The Indian goes at pleasure from one to the other.

† Two sick people, a man and his wife, whose end seemed to be fast approaching, being unable to undertake the journey,

sion of these journeys, which are undertaken in companies, apostates fall off, and new converts come in. The first, some of whom become the bitterest enemies to the Spaniards, the missionaries endeavour, on their excursions, to regain by gentle means; and if they do not succeed, they have recourse to the armed force. Hence many of the hostile events between the Spaniards and Indians.

The Indians die in the missions, in an alarming and increasing proportion. San Francisco contains about a thousand Indians: the number of deaths, in the last year, exceeded three hundred; it amounts already this year, (till October,) to two hundred and seventy, of which forty occurred during the last month. But the number of proselytes must exceed that of the apostates and the excess of deaths. Five missions were named to us, which have been founded in this province, since the time of Vancouver. On the other hand, several of the missions of the Dominicans in Old

---

had remained behind the throng of departing Indians. They did not return to the mission; they laid themselves naked as they were on the damp ground, on the shore near our tents, without a covering from the stormy rainy nights. Their looks were fixed on their blue mountains; they saw their native home; and thus consoled themselves for not being able to reach it. The fathers after a few days, observing them, sent them back into the mission, addressing them with much mildness.

California, have ceased to exist; and the converted people may be considered as nearly extinct.

There is no medical assistance here, except bleeding, which is said to have been taught them by a ship's surgeon; and this remedy being since applied on every occasion, is more fatal than advantageous. Particularly one disorder, which, though the opinions are divided, has probably been spread by the Europeans, carries off its victim without opposition. It likewise prevails among the savage tribes: these latter do not, however, disappear from the earth with the same dreadful rapidity. The number of whites, on the other hand, increases.

The contempt which the missionaries have for the people, to whom they are sent, seems to us, considering their pious occupation, a very unfortunate circumstance. None of them appear to have troubled themselves about their history, customs, religions, or languages: "They are irrational savages, and nothing more can be said of them. Who would trouble himself with their stupidity? who would spend his time upon it?"

In fact, these tribes are far below those on the north coast, and the interior of America. In their general appearance, they resemble each other, except the Tcholovonians, whom we soon learnt to distinguish by their marked physiognomy, which the fathers could not do. They have all a very savage look, and are of a very dark colour. Their flat, broad countenance, with large staring eyes, is shaded by black, thick, long, and smooth hair. The gradations of colour, the languages, which are radically different from each other; the mode of life, arts, arms, in some of them various lines tattooed about the chin and neck, the way in which they paint themselves for war and for the dance, distinguish the different tribes. They live among the Spaniards, and among themselves in different, friendly, or hostile relations. Among many of them their arms consist of bows and arrows; some of these are of extraordinary elegance, the bows light and strong, and covered with the sinews of animals on the convex side; among others it is merely of wood, and rudely made: some possess the art (women's work) of constructing neat and water-proof vessels of coloured blades of grass; but the Indian, for the most part, forgets his industry in the missions. They all go naked. They do not possess horses nor canoes of any kind; they only know how to fasten together bundles of rushes, which carry them over the water by their comparative lightness. Those who live near rivers subsist principally on salmon, which they catch in baskets; those in the mountains on wild fruits and grain. They neither sow nor reap, but burn their meadows from time to time to increase their fertility.

The South Sea islands, far distant from each other, and dispersed over nearly one-third of the torrid zone, speak one language. In America, here in New California, tribes of one race, living near to each other, speak quite different languages. Every fragment of the history of man is of importance.

We must leave it to our successors, as our predecessors have done to us, to collect more satisfactory information respecting the natives of California and their languages.* We had proposed this as our object in a journey which we intended to make to some of the nearest missions. Business however of another kind kept us at San Francisco, and the period fixed for our departure came without our being able to afford time for this journey; for the rest we refer to the accounts of La Peyrouse and Vancouver, which we found very correct. Since their time there has been but little change in California. A fort, erected in a good situation, guards the harbour of San Francisco. The Presidio is new built with stone, and covered with tiles. The building of the chapel has not been begun. In the missions they build in the same manner, and the barracks of the Indians at San Francisco are of similar construction. An artillerist has erected mills in the missions, worked by horses; but they are now for the most part out of order, and cannot be repaired. At San Francisco is a stone which a horse turns without mechanism over another stone, the only mill in order. The Indian women rub the corn between two stones for immediate use. A windmill of the Russian American Company's settlement creates astonishment, but does not find imitators.

* De Lamon has given valuable information in La Peyrouse's Voyage, on the languages of the Achastlians, and Eclemaches, near Monterey. For further particulars, see Adelunge, Mithridates, 3. 3. p. 182.

Some years ago, when artizans were brought here at a great expense to teach the necessary arts, the Indians profited more by their instructions than the *gente rational* (rational people) as the Spaniards call themselves.

We observed with regret, that the best understanding does not exist between the missions and the Presidio. The fathers consider themselves as the first in this country, and the Presidios merely sent for its protection. A soldier, who constantly carries and often uses arms, unwillingly bears the government of the church. The Presidios, living only on their pay, depend for the supply of their wants upon the missions *, from which they purchase for ready money; they suffered distress in this latter period, neglected by the mother country, and accused the

* At the head of each mission are two Franciscans, who have engaged to spend ten years in this hemisphere. They are dispensed from the rules of their order, and receive four hundred piastres each from the crown. Several missions are under one Presidio. The commandant of the Presidio, captain of a company, has under him an artillery officer, a commissary, a lieutenant, an ensign, and eighty men; each of whom receives two hundred piastres a year. The Spaniard is always on horseback. Horses and oxen are kept in herds, and are almost wild; when wanted, they are caught with a *lazo* (a noose). The arms are, the lance, shield, and musket. The Presidios have no tillage; the officers hardly cultivate a little garden-ground; they consider themselves as exiles, who wait with impatience their speedy recal. The Pueblas, as they are called, which are few in number, are Spanish villages. Some colonists, and veteran soldiers constitute the population. Their wives are, for the most part, Indian women. The Governor of New California, at Monterey, as well as the Governor of Old California, at Loretto, is under the Viceroy of Mexico.

mission of not endeavouring to relieve them. Before we conclude, we must mention the generous hospitality with which both the military and the missionaries strove to supply our wants, and they willingly granted an unconstrained freedom, which we here enjoyed on Spanish ground. We dedicate these lines of remembrance and gratitude to our friends in California.

The following races of Californians were named to us, as living within the precincts of the mission of San Francisco.

The Guymen Utschiun Olumpali Soclan, and Sonomi	Speak all one language; they are the most numerous of any in the mission of St. Francisco.
The Chulpun Umpin Kosmitas Bolbones Tchalabones Pitem Lamam Apalamu and Tcholovones	Live at Rio del Sacramento; and all speak but one language. They are the most skilled in arms. The Tcholovones, a war-like tribe, are united with the Spaniards against the other Indians.
The Suysum Numpali and Tamal	Tattoo themselves, all speak the same language, and live to the north: the Tamal live towards the north-west.

The Ululato live more to the north than the Suysum, and only a few come into the mission.

THE PHILIPPINE ISLANDS.

Cavite, situated on the extreme point of a tongue of land, which projects into the beautiful and much-frequented bay of Manilla, and cuts off a part of it, is the most unfavourable station for a traveller, who, during his short stay in Luçon, intends to employ his time in examining the nature of the country. The tongue of land, and the finely-cultivated banks of the bay up to Manilla, belong to civilized man. Between the houses and villages you see only rice-fields, gardens, and plantations, in which grow the plants of both the Indies.

We had an opportunity to make only one excursion, of eight days, into the interior, to Taal, and the volcano, of the same name, in the Laguna de Bonborig. The military escort accompanying us, which was a mark of Spanish pomp, was very troublesome, and increased the expenses of a journey where only a guide would have been requisite among the mild and hospitable Tagalese. The island of Luçon is every where high and mountainous; the highest summits do not seem, however, to exceed the woody region. Three volcanoes rise from it: first, in the north, the Aringuay, in the territory of the Ygorrotes, in the province of Ilocos, which, on the 4th of January, 1641, broke out at the same time with the volcano of Iolo, and the Sanguil, in the south of Magindanao, on which occasion this island presented one of the most terrible scenes recorded in history*; the noise was heard on the continent of Cochin-China. Secondly, the volcano de Taal, which particularly threatens the capital, from which it is distant a day's journey; and, lastly, the far-seen Mayon, near the Embocadera de San Bernardino, between Albay and Camarines.

Gold, iron, and copper-mines, which are very rich, but neglected, show that there are other mountains as well as volcanic. On the way we went, we saw no other than volcanic tuff, consisting of ashes, pumice-stones, and dross; and, in Manilla, Cavite, Taal, Balayan, &c., no other stone for building but this same tuff and calcareous reef-stone, procured from the sea. The granite, used in Manilla for building, is brought here as ballast from the coast of China.

As you go from Cavíte, southward towards Taal, the land insensibly and gradually rises till you reach the eminences on the other side, which are rugged and steep, and from which you may overlook, at your feet, the Laguna de Bongborig, and the large smoking crater, which forms in it a dreary, naked island.

* The journals of Manilla mention the destructive earthquakes, in the year 1645 and 1648.

The lake (the Laguna) is about six German miles in circumference; it empties itself into the Chinese Sea by an outlet, navigable now only for small boats, though formerly it could carry larger vessels; it runs with great rapidity, and the length of its course is above a German mile. Since the devastation in 1754, Taal has been removed to its mouth.

The water in the Laguna is brackish; but it is, however, drinkable. In the middle it is reported to be unfathomable. It is said to be full of sharks and caymans, of which, however, we saw none.

As we wère embarking from the Laguna for the island, the Tagalese exhorted us to look round us in this haunted place, but to keep silence, and not to irritate the spirit by any incautious, inconsiderate word. The volcano, they said, showed symptoms of displeasure whenever a Spaniard visited it, and was indifferent only to the natives.

The island is nothing but a mass of ashes and scoriæ, which has fallen in itself, and formed the wide irregular crater, which creates so much terror. It does not appear that lava has ever flowed out of it. From the bank, where a little grass grows in scanty spots, and where some cattle are kept to pasture, you climb, on the east side, up a bare and steep ascent, and, in about a quarter of an hour, reach the edge, from which you look down into the abyss as into the area of an extensive circus. A pool of yellow, sulphureous water occupies about two thirds of the bottom. Its level seems to be the same as that of the Laguna. On the southern edge of this pool are several hills of sulphur, which are slowly burning. Towards the south and east of it, a narrower crater is beginning to form itself in the interior of the great crater. The arch which it makes surrounds, like the *moraine* of a glazier, the burning hills by which it is produced, and rests with both its ends on the pool. The pool boils, from time to time, at the foot of the burning hills.

You can clearly distinguish, in the internal wall of the crater, the situation of the differently coloured scoriæ of which it consists. Smoke ascends from some points of it.

We observed from the place where we made a drawing of the crater, a place on the opposite side of it, where a fall into the interior seemed to afford a slope, from which it might be possible to descend to the bottom. It cost us much time and trouble to gain this point, as we found the sharp and pointed edge on which we walked, in many places impassable, and were frequently obliged to descend on the outside almost to the bank. Being under the wind of the fire, we were but slightly incommoded by the sulphureous exhalations.

The place just mentioned is that on which, during the last eruptions, the water poured that was thrown up. We attempted to descend into several clefts, but were ultimately obliged to abandon our intention after we had reached about two-thirds of the depth. We were not provided in Taal with the cords we required, and by the assistance of which we might probably have descended the perpendicular wall of several fathoms high, which first presented itself to us, without being able to reach the bottom, as the precipice became always steeper the farther we descended. We found, in this neighbourhood, the ground covered with crystallized salt. * The time was too short to permit us to visit other hills. The other craters are at the foot of the principal crater.

The most terrible eruption of the Volcano de Taal was in the year 1754. Its devastating progress is circumstantially related in the twelfth chapter of the thirteenth part of the history by Fr Juan de la Conception. The mountain was tranquil after the former eruptions, (the last took place in the year 1716,) and sulphur was obtained from the apparently extinguished crater. It began to smoke anew in the beginning of August; and, on the 7th, flames were seen, and the earth trembled. The consternation increased from the 3d of November to the 12th of December; ashes, sand, mud, fire, and water were thrown up. Darkness, hurricanes, thunder and lightning, subterraneous roarings, and long-protracted, violent, and repeated earth-

* According to Dr. Mitscherlich's examination. *Plumose alum.*

quakes, alternated in frightful succession. Taal, lying at that time on the banks of the Laguna, and several villages, were totally ruined and overthrown. The mouth of the volcano was too confined for such eruptions; it widened considerably, and a second opened, which likewise threw up fire and mud. Nay, even more, the fire broke out in several places in the Laguna, at a considerable depth below the surface of the water, which boiled up. The earth opened in many places, and a deep gulf yawned particularly wide, extending far in the direction to Calanbong. The mountain continued to smoke a long time. There have since been eruptions, though with decreasing violence.

The beautiful forests which clothe, in luxuriant verdure, the mountains and a part of the country, extend to the sea, into which the mangroves and other trees hang down. We have had but a cursory view of these forests, in a trodden path, and have not penetrated far enough into them, to be able to give a proper description. The fig-tree seemed to us to predominate. Several species rest themselves, as mighty trees, on a singular net of stems, and creeping plants, which twine round the rocks, and spread themselves out over them. Others rise with slender stems to an astonishing height, and you perceive the enigmatical fruit on the under branches of the trees, whose crown is lost amid the verdant canopy of the forest. Some species remain shrubby, and others twine. We missed in the forests the beautiful form of the acacia-trees, with their variously feathered leaves. The numerous species of the siliquose plants adopt here all imaginable forms. The fern-plants, and particularly the arborescent kinds, the *Lianes*, the *Orchideæ*, plants which, in Brazil, form airy gardens on the tops of the trees, seem to be very few in number, or to be entirely wanting, like the *Cactus* or the *Bromeliaceæ*. Nature bears a different and more tranquil character. The kinds of palm are more numerous than at Saint Catharine's. Several of these are insignificant, the slender prickly calamus is probably the most remarkable. Among the *Aroideæ*, the *Pothos scandens*, with its grass-like leaves, narrowed in the middle, creeping up the branches of the trees, is very striking by its form.

On the low grounds and banks of the rivulets grows the elegant bamboo-cane *, whose slender halms shooting forth in thick bushes, lightly glide over each other, sounding to the play of the wind, and a close thicket offers the richest variety of plants.

On the plains, forests alternate with savannahs,

* The halm of the bamboo shoots up, in one rainy season, to the greatest height it can attain, turns woody the following year, and spreads out side-branches without growing higher. The young shoot is eatable, like that of the asparagus. Some of the kinds described by Loureiro are native here; we saw the blossom of none.

whose Flora is extremely scanty. Two kinds of grass, the halms of which reach a height of about eight feet, and which are scorched by the sun, appear to have been sown, and ripening for the harvest. A very few dwarf-plants, for the most part siliquose, conceal themselves in their shade, and an arborescent *Bauhinia* rises here and there among them.

These savannahs are set fire to, either to prepare them for cultivation, or to procure grass of a younger growth for the herds. The fire crackles over them, and a small species of hawk and other birds actively fly round the clouds of smoke that rise before the advancing fire, in chase of the insects which fly to escape from it.

Circumstances confined our researches in the system of organized nature, almost exclusively to botany and entomology. We, however, find occasion to say a few words on a marine insect, less known to the learned than the mercantile world.

Under the general name of *Biche de mer*, in Malayan, *Trepang*, and in Spanish, *Balate*, certain dried and smoked *Holothuria*, of seven, and perhaps more kinds, are brought to the market of Canton, each having its particular value and name. The same epicurism of the Chinese, which attaches a high value to the birds' nests which are known even in Europe, maintains also the value of the *Trepang* by the great demand. The Malays look for it as far as the coast of New Holland, in the gulf of

Carpentaria; the Malays and Chinese, as far as the coast of New Guinea; the English have it collected in the Pelew islands by sailors left there for the purpose. The Spaniards bring it from the Marianes, and, as it gradually vanishes from the coasts on which it is sought, they undertake voyages to the Carolinas, of which we shall speak in another place, to discover it. The *Trepang* appears to be also collected in the Indian Ocean, and epecially on the island of Mauritius, for trade. These *Holothuria* are particularly found on the coral reefs, where some species are gathered, as at Radack, on the beach when the tide is down, whereas others seem to live in deeper water. We have had an opportunity of examining more accurately, and of drawing this one species. It is one of the smaller and less valued; the others resemble it. All real *Holothuria* may be eaten as *Trepang*. This precious worm is collected in many places in the Philippine islands.

The entomological kingdom is rich in these islands. The butterflies, beetles, and bugs are particularly beautiful. A scorpion seems to be the same kind as that met with in the islands of the Great Ocean, and which we also collected in Radack, but we found them much larger here. Termites and musquitoes are the torment of the inhabitants. A large *Mantis* (leaf-fly), which is very frequent in Manilla, probably gave occasion to Pigafetta's story of the living leaves on a tree in the island of Cimbonbon.

This tradition, and similar ones of the living sea-weed, &c. of the people with tails, &c. which Fr. Juan de la Conception relates in his history are still recorded by the Spaniards; for nobody here has any taste for natural history, or indeed for any science, and every one asks only for what he wants, or what is necessary to him in his business. The collection of natural history of D. Gonzales de Caragual, intendant of the Philippine islands at the time of La Peyrouse, (1787,) has since been sent from Manilla to the mother-country.

The learned Cuellar, who was sent from Spain with a commission to promote several economical objects, the cultivation of cotton, of cinnamon, &c., and, after a long residence on these islands, died several years ago at Manilla, had formed a botanical-garden in Cavite, of which there is now not a trace remaining. Cuellar sent specimens of natural productions of all kinds to Madrid; took much pains to purchase Chinese books; enriched the gardens of Madrid and Mexico with the seeds of several plants of this country, and maintained learned correspondences in both the Old and the New Worlds. We have examined his papers, and convinced ourselves, that every thing relating to science had been snatched from oblivion and sent to Spain. It appears that Cavanille has described his collection of plants, as well as those of Malespina's expedition, which lost here one of its naturalists.

To collect the abundant harvest which natural science has still to reap here, requires a longer stay, and excursions to the different, and, more particularly to the more promising southern islands and into the interior of them. There is still much to be done, and employment for many.

The Philippine islands have to boast of more minute historians than many European kingdoms.* We are obliged to the translator of Zuñiga

* Antonia de Morga, Sucesos de Philipinas, Mexico, 1603. Pedro Murillo Valarde, Historia de la Provincia de Philipinas de la Compañia de Jesus, Manila en la imprenta de la Comp. de Jesus, 1749, 2 vols. fol.

Juan de la Concepcion, Recoleto Augustino descalzo. Historia General de Philipinas, Manila, 1788—92. 14 vols. 4to. Joaquin Martinez de Zuñiga del ordende San Augustin Historia de las Islas Philipinas, Sampaloc, 1803, 1 vol. 4to: of which an English translation has already passed through two editions. An Historical View of the Philippine Islands, from the Spanish of Martinez de Zuñiga, by John Maver, London, 1814.

Poblacion de Philipans, fol.: an imperfect statistical table, with many errata in the figures, printed at Cavite en S. Telmo, 1817. It appears, that similar tables have appeared previously, from about the year 1734.

Carta edificante o Viage a la Provincia de Taal y Balayan, por el Abate Don Pedro Andres de Castro y Amcedo, 1790, 4to.: MSS. in our possession.

Besides these, the following historians are mentioned, which we have not had an opportunity to consult.

Fr. Gaspar de san Augustin.

Colin Historia de Philipinas: an extract from the following, Pedro Chirino, Historia de Philipinas, 1 vol., fol.: MSS. of the library of the Colegio, and different chronicles and histories of several monks' orders, or rather of their province, the Philippine for having spared us the duty of stopping at this disgusting history, which consists of nothing but a mixture of monastical disputes, and of the struggles of the spiritual power with the temporal; in which the accounts of the missions in China, Japan, &c. appear in a very unfavourable light. Fr. Juan de la Conception has brought the history down to the government of Aranda, in the year 1764. We shall now take a cursory view of the present state of this Spanish settlement.

The Spaniards include under the jurisdiction of this government, the Mariana islands, the Caroline islands, of which wrecked boats from thence gave them the first information, and to which they resolved to extend their faith and their yoke; and, lastly, the southern islands of the Philippines, Magin-

---

islands, which has been preserved in MS. in the convents of these orders at Manilla.

History of the Marianes.

Charles Gobien; Histoire des Isles Marianes, nouvellement converties à la Religion Chrétienne, et de la Mort glorieuse des Premiers Missionaires qui y ont prêché la foi, Paris, 1700.

History of the Discovery of the Caroline Islands, and the Missions to them.

Lettres édifiantes, v. 1. 2d edition, v. 11. 16. 18. Murillo Velarde and Juan de la Conception, seem to have consulted no other authorities than the letters and reports here enumerated.

On the Pelew Islands in particular.

George Keate, Esq. An Áccount of the Pelew Islands, from the Journal and Communications of Captain Henry Wilson, 5th edition, London, 1803, 4to.

danao, Jolo, &c. possessions of their hereditary enemies, the Moors, or Mahometan Indians, who do not cease to spread consternation and devastation, by piracies, over all the coasts in the hands of the Christians.

The Presidio of Sanboangan, on the west point of Magindanao, is designed to keep this race in check, but is, in fact, like the government of the Mariana islands; only a source of revenue to the commandant, calculated, according to the years of his government, to enrich himself by the exclusive trade with all the salaries appointed for the garrison and public officers.

The expeditions which are sent out from Manilla in armed boats, are not more serviceable. They only protect the smuggling trade, and Christians and Moors avoid each other with the same diligence. Only the Bay of Manilla, which was represented as unsafe by La Peyrouse, seems to be now closed against the pirates.

There are, in the Philippine islands, besides the Spaniards, who are looked upon as foreign masters, and the Chinese, their parasites, two native races of men: Papuas in the interior, and the Malays in a more extended signification, or south-landers on the coasts. The Spaniards are but few in number. The Chinese, who are called Sangalese, that is "Wandering Merchants," (they might be called the Jews of this part of the world,) differ in their numbers at different times; sometimes greater, and at other times less. Their civil relation does not rest on a firm, fixed compact; and history shows them to us sometimes as tolerated, sometimes as persecuted, and at other times as insurgents. Many of them, to settle with more security, suffer themselves to be baptized, and when they leave Manilla, with the riches they have acquired, in ships of their own nation, frequently send to the archbishop the white Neophytes' dress and cross, which they have received from him, that he may confer them on others of their countrymen.

The Papuas, the first possessors of the soil, the Aetos or Negritos of the Spaniards, are savages, who, without a fixed abode, without agriculture, rove about the mountains, and live by the chase, on wild fruits and honey. They cannot be enticed to adopt any other mode of life. Even those who have been brought up from their childhood among Spaniards are wavering Christians, and not seldom fly from their patrons to the people of their own colour in the desert. They appear to be more hostile to the Indians who drove them out, than to the Spaniards who are their avengers. Very little is known respecting them, and we were not fortunate enough to obtain any decisive information on this subject. They are, in general, represented to be a mild and unsuspicious people; and, in particular, have never been accused of eating human flesh. They go naked, except an apron of the bark of trees. We tried in vain to see this article of dress, or any thing of their manufacture, and must leave it undetermined, whether this bark is worked raw, or after the manner of the southern stuffs. We saw of this race only two young girls, who were brought up in Manilla and Cavite, in Spanish families. There were, besides, two men condemned to work at the fortifications in Cavite.

Of the Malays, the Indios of the Spaniards, there are different tribes, and people speaking different languages, whom history makes to come from Borneo and Magindanao. Many tribes in the interior have retained their liberty: the inhabitants of the coasts are Christians, in the hands of the monks, and subject to the Spanish crown.

The independent tribes deserved our most particular attention, but we were unable to procure any farther information respecting them. They differ from each other in many points, and what is applicable to one is not to be extended to all. It is to be observed that, by some of them, chastity is held in great honour, not only among the women, but also among the young girls, and is protected by very severe laws. A kind of circumcision is said to be a primitive custom among others, and not to be derived from Mahometanism.

The Indíans of the Philippines are in general a friendly, harmless, cheerful, and cleanly people, whose character reminds us more of the inhabitants of the eastern islands than of the real Malays, or the cruel Battas. Corruption of manners prevails only among the lowest class, who surround strangers in Manilla and Cavite. We refer to the authorities quoted, and to Pigafetta's voyage, for the manners, customs, and numerous superstitions of these people. The table of population for the year 1815 estimates the subjects of Spain, in the jurisdiction of this government, at two millions and a half.* The receiving of baptism is considered as a mark of subjection. In this number are not included two thousand families of unconverted Indians (Tinguianes) of the province de Ylocos, in the north of Luçon; about a thousand families of the unconverted Indians (Ygorrotes †) in the mountain of the same province; twelve hundred families of the Negritos of the same mountain; and, lastly, above nine hundred families of the unconverted Indians of the province of Calamianes, who all pay tribute, each in different goods, and particularly the Negritos, in virgin wax. The population of Manilla is estimated at nine thousand souls, exclusive of the

* The usual manner of making the census, is by the tribute which is paid by every family. Tribute or family is reckoned on an average at five souls. In the same table it is mentioned, that the population has increased about one million seven thousand souls, since the year 1734.

† The physiognomy of these Ygorrotes de Ylocos, and their lighter colour, prove that they have mixed with the companions of Limahon, who fled to their mountains, when Juan de Salceda besieged the Chinese in Pangasinon.

clergy, garrison, settled Spaniards, Europeans, and Chinese, who make from four to six thousand.

Manilla, with its harbour of Cavite, appears to be the only considerable Spanish town in the Philippines. In the provinces, you see the splendid edifices and temples of the clergy rising among the slight and cleanly huts of the natives, which, as at the time of Pigafetta, are raised on posts made of bamboo-cane and prickly interwoven calamus, and covered with Nepa leaves, and may be compared to elegant bird-cages. The fire often consumes such villages as easily and rapidly as the dry grass of the savannahs; and, in a few days, they rise again from their ashes.

The Spaniards in Manilla live chiefly in that part of the town which is properly fortified, on the left bank of the river. The suburbs of the Chinese, surrounded with shops and booths, and that of those Tagalese with beautiful gardens, extend on the right bank. The streets of the town are regular. The houses massive, of one story, built on a ground-story, which is not used. The dampness of the rainy season has taught the Spaniards to follow the example of the natives. They are entirely surrounded on the outside with galleries, the windows of which, instead of glass, are furnished with transparent shells. The spacious, airy and shady rooms afford a pleasant retreat against the heat. The architecture of the convents and churches, which compose the chief edifices of the town, is not bad. On account of the earthquakes the walls are of an extraordinary thickness, and secured by beams worked in. Some of these churches possess pictures by good masters; several altars are ornamented with wooden statues, which are not without value as works of art, and are the work of the Indians. But any thing done by Indians is not valued here. We employed the few cursory hours we spent in Manilla chiefly in the convents, where we hoped to obtain information on important subjects. In the seminaries of the Chinese and Japanese missions we did not find one monk who was conversant in the sciences and literature of these people. The strangers learn at the place of their destination itself the necessary languages; and what you ask for in the pretty large libraries of Manilla is precisely what they want, namely, the department of the philology and literature of the inland languages, and of the nations whom they desire to convert to their faith by the missionaries sent thither.

The inquisition appears to slumber now, but the habit of caution still remains, and you may see that people are uncomfortable, and dread it as a spectre they do not see.

The Spaniards living here display great luxury. The equipages are numerous and elegant. The profusion of their dishes, the number of meals they give on one day, almost causes satiety. The object of every one is to amass riches, and a well-known

Spanish proverb, says, "I did not come to India merely for a change of air."

More extended liberty will make the trade of Manilla flourishing; and the restraints to which it is subjected in Canton may cause the market between China and the rest of the world to be removed hither. Every one trades, and the monks, who possess the ready money, are willing to confide it to speculators for certain profits, and for certain enterprizes, in the risks of which they partake. Sugar and indigo seem to be at present the principal articles which are sought for the European markets; cotton and stuffs of native manufacture are exported to Mexico. The Chinese buy *Trepang* and birds' nests. The shell which, in many parts of India, passes as coin, and which these islands furnish, pearls, mother of pearl, amber, &c. can scarcely be taken into the account. These islands might furnish many more productions for trade than they really do; the coffee, which is of a particularly excellent quality, is, like the cocoa, cultivated only for home consumption. The cinnamon, which is said to grow wild in many parts of the forests, sago, &c. seem not yet to have become sources of wealth.

When history shall have sealed the separation of the two Americas from the mother-country, the Philippine islands will yet remain to the Spanish crown, and, by a wiser administration, may compensate her for the loss of an immense territory from which she did not know how to derive the advantages which it offered.

The Indians are proprietors and free subjects, and are treated as such. The forts, which are built in every place on the coast against the Moors, are in their power, and are garrisoned by them. The privileges of their noble families have fallen into desuetude, every district, every village chooses its chief, and the choice is only confirmed. These Governadorcillos, Capitanos, &c. who are addressed by the Spaniards with the title of Don, possess all the legitimate authority; but the consideration, the riches, and power, are all on the side of the monks. The fathers, who govern the people, drain them in various ways, and after the church dues have been paid and the priests kept the best for themselves, the poor man parts with his last saving to buy scapularies and images of saints.

The tribute which is paid to the king, is but a reasonable burthen; but the administration of tobacco, which is one of the principal necessaries of life, without any distinction of age or sex, is very oppressive. The fields, which the people formerly cultivated for their own account, now lie fallow. The Indian fears that every new production will bring upon him a new tax. Only a small duty is paid for the areca-palm, the nut of which is chewed with the betel-leaf and lime.

The food of the people is rice; and, with this, all the fruits with which nature has so lavishly endowed this fertile soil, and among which we will only mention the highly-extolled mango *, two sorts of bread-fruit trees, the common one of the South Sea islands, and that peculiar to the Philippines, the plantain and the cocoa.

The domestic animals originally in this archipelago, were the hog, the goat, the dog, the cat; fowls, the goose, and according to Zuñiga, also the *Caraboa* or East Indian buffalo†, which must be distinguished from that of the south of Europe, and for which we refer to Marsden's account.‡ The *Caraboa* is found in the mountains, either in a savage state or run wild. The Spaniards first introduced our kinds of oxen, the horse, and sheep.

The cock-fight, already mentioned by Pigafetta, is the greatest diversion of the Indians. A good fighting-cock is the pride and delight of his master, who carries him about everywhere in his arms.

* Zuñiga doubts whether the mango was originally native, or whether the Spaniards brought it from the coast of the continent. The same writer most unaccountably reckons the sugar-cane among the plants brought here by the Spaniards. Pigafetta expressly mentions the sugar-cane in Zebu. Don San Jago de Echaparre has tried in vain to naturalize the walnut and chesnut-tree. He sowed both kinds, but without effect, several times on mountains of the interior, and on the skirts of the forests.

† Pigafetta does not seem to have met with the *caraboa*, in the islands of this archipelago, which he visited. He mentions only the buffalo in Borneo, with the elephant and horse. The word *Caraboa*, *Karbare*, is Malayan.

‡ Marsden's Sumatra, p. 94. the first edition.

He is tied in the dwelling-house by the foot, and attended with the utmost care. Their love of fighting, and the courage of these animals, is caused by the abstemiousness to which they are subjected.

The palm-wine, or rather brandy, is now, as it was in the time of Pigafetta, the favourite beverage of the Indians. We find the manner of procuring it first described by Marco Polo. The flower spatha of the cocoa-palm, before it opens, is tied together, the point cut off, and a vessel of bamboo fastened to it, which receives the juice as it issues out. This juice is collected twice a-day, and when such a source dries up, another spatha ripens on the same tree to supply its place. From this juice, which, when taken fresh, is very cooling, a different process produces wine, vinegar, brandy, or treacle.* Many cocoa-trees become evidently unfruitful by too luxuriant growth; to avoid which, it is customary to make deep incisions in the trunk. But when a tree has in this manner become useless, it is felled, and the cabbage or unfolded leaves in the middle of the crown afford a very pleasant dish.†

A particular kind of musa (*Pisang*, *Banana*),

* The sweet treacle of the Pelew islands is procured in this manner, only from the cocoa-palm. Fermented or distilled liquor seems not to have been introduced there.

† We have particularly observed, in Guahon, the unfruitfulness, or, according to the Spanish expression, running mad (*tornar loco*) of the cocoa-tree, and the remedies employed against it.

which bears a small eatable fruit, is cultivated on account of the flax, which is procured from its trunk, and which seems to deserve the preference to many others. The fibres are of the whole length of the stalk (about eight feet), and of different fineness, according to their exterior or interior situation; so that the same plant furnishes the flax from which the excellent cables are made, which are here, for the most part, used by the Spanish marine, and from which the fine-striped stuff is woven for the neat shirts, which are part of the costume of these cleanly people.*

A palm-tree (*Palma de Cabello negro*) furnishes a strong black bark, which is also used for manufacturing cords and cables. (The Chinese cables braided of the prickly calamus, which many navigators of the Great Ocean are obliged to use, is said to be the worst and most unsafe.) This palm-tree is cultivated and multiplied on account of its great use.

Lastly, the bamboo and prickly calamus must be mentioned as the most serviceable plants in this hemisphere.

The Tagalese with his *bolo* (a knife, which he constantly carries by his side, well ground, and which is the only instrument he uses in all his

* The Carolinians also prepare their stuffs, resembling mats, from the fibres of the musa, which according to Kadu's account, is cut down for this purpose before it bears fruit. Do they also possess the species above mentioned?

mechanical works, and which serves him at the same time as a weapon), builds himself his house of bamboo and prickly calamus, and provides it with all the necessary utensils and vessels. The soil affords him meat and drink, stuff for clothing, tobacco, the areca-nut, and betel for his enjoyment. A fighting-cock makes him happy. Nature is here so bountiful; man so contented! He requires so little for his sustenance and joys, and yet this little he can not always procure!

THE MARIANA ISLANDS.—GUAHON.

The Mariana islands form a volcanic chain, lying in a direction from north to south. The volcanoes and the seat of the subterraneous fire are in the north of the chain, where barren, burnt rocks are enumerated among the islands.

In Guahon, the most southern of them, and, at the same time, the largest and most considerable, only slight shocks of earthquakes are felt. Guahon appears, from the N.E. side, a tolerably high level land, the shores of which are rugged precipices. The neighbourhood of the eminences and town bear a different character, and have lofty hills and beautiful vallies.

We found no other kind of rock but madrepores, calcareous spar, and lime-stone.

The island is well wooded; the Flora appears to be rich, and the vegetation luxuriant. The forests descend on the sloping shores to the sea; and different kinds of *Rhizophora* bathe, in covered places, their foliage in the waves. Nothing can equal the aromatic odour which wafted to us on our arrival over the surf, when we were seeking an anchorage. The orange-trees, like other fruit-trees of various kinds, are grown wild; memorials of a former more flourishing cultivation. Many foreign plants have amply increased the Flora; for example, the prickly *Limonia trifoliata*, which cannot now be checked, and the *Indigofera tinctoria*, which nobody understands how to turn to advantage. The bread-fruit tree, the cocoa, the plantain, are here in abundance; the *Magnifera indica* has been planted, but has not yet become naturalized. We found only the different kinds of plants which are common on the continent of Asia, and the islands of the Great Ocean; for example, the *Baringtonia speciosa*, and the *Casuarina equisetifolia*. We missed the following kinds, growing in New Holland, the *Proteaceæ*, *Epæridéæ*, *Myrtoideæ*, and acacia, with simple leaves. We met here with most of the plants growing in Radack, of which we afterwards missed several in Luçon, for example, the *Tacca pinnatifida*, which, though naturalized and planted in Cochin China, seems to be wanting in Manilla. There are two different sorts of pandanus, and several fig-trees.

Besides bats, (we found the *Vampyrus*:) the only original native quadruped is the rat, which is so generally spread over all the islands of the South Sea. The Spaniards have introduced, besides our common domestic animals, of which we did not find any there, the guanaco from Peru, and a stag from the Philippines; the stag, during the time of the Governor D. Thomas. Many of these animals have now run wild in several of the islands. There are some different kinds of land-birds, and, among others, a hawk. We observed among the amphibia, an iguan, and a large sea-turtle; among the zoophytes, some of the *Holuthuria* species, which furnish, under the name of *Trepang* (*Biche de mer*, *balate*), such an important article in the trade with China.

The melancholy history of the Marianas is sufficiently known in Europe. We refer to the Histoire des Isles Marianes nouvellement converties à la Religion Chrétienne et de la Mort Glorieuse des Prèmiers Missionaires qui y ont prêché la Foi, par le Père Charles Gobien, Paris, 1700; and to the extract of it in Burney's Chronological History, vol. iii. p. 271.

These islands were discovered by Magellan; they were called, by the natives, Laguas; the Spaniards called them Las Islas de los Ladrones, et las Velas Latinas; and, lastly, Marianas. The pious missionary, Don Diego Luis de San Vitores, landed at Guahon in the year 1667; he wished to bring the people salvation, but he was followed by soldiers and arms. Already, before the end of the century, the work was finished, and this nation had ceased to exist! The Spaniards call it *Pacificar*.

"This considerable decrease comes from the subjection to which they were compelled by arms. Fond of liberty, they could not bear a foreign yoke; and being unable to shake it off, they either hung themselves, or deprived themselves of life in another manner. The women purposely procured abortion, and threw their own offspring into the water, being convinced that, by this premature death, which would deliver them from misery and trouble, they conferred happiness and salvation upon them. Thus they considered dependance as the greatest and most intolerable wretchedness. An epidemic disorder also contributed; which, in the beginning of the century, carried off almost all that remained." — Fra. Juan de la Conception, Historia de Philippinas, t. vii. p. 348.

Don Pedro Murillo Velarde gives the same picture, with the same circumstances. We willingly leave it to the Spaniards to speak for themselves.

The original population, according to Fra. Juan de la Conception, was 40,000; according to Murillo Velarde, 44,000. It is said, in the Nouveau Voyage à la Mer du Sud (Marion), that the population formerly above 60,000, had decreased to 8 or 900. The remnant of the natives were collected on the islands of Guahon and Saypan, in 1695; and, on the sickness which broke out immediately after, on the former island alone. After the census, without any date of the year, which Murillo Velarde (published in Manilla, 1749,) communicates as the latest account, there were 1738 inhabitants. The population had increased, in the year 1783, to about 3231; and, in 1816, to 5389 souls. *

* It must not be forgotten, that, in former times, hundreds of Philippine islanders were brought to Guahon, to increase the mission; and that their descendants are reckoned in these census.

But the Christian descendants of those who escaped the ruin of their race, and survived their independence, have lost all the peculiarities of their ancestors, all their arts, and, for the most part, forgotten their language.

Gobien appears to be the first who made the ridiculous assertion, that the inhabitants of the Mariana islands were first made acquainted with fire by Europeans. The historians of Manilla repeat this circumstance, and Velarde applies to them the *Nulla Getis toto gens truculentior orbe;* and we are astonished to see able authors, from whom a sounder judgment might have been expected, lightly led into unpardonable errors.*

This people belongs to the family who, related by character, customs, and arts, and connected by commerce and navigation, inhabit the islands

* Burney shows here, too, how the most solid learning is with him in good hands, l. c. p. 312. How could the inhabitants of the islands, on which many volcanoes burn, be ignorant of fire? Pigafetta enumerates, among their food, the flesh of birds, without remarking that it was eaten raw. We observe, *en passant,* that the sow, which, according to this voyage, Magellan had killed on his arrival at Huna-rura, in the Philippine islands, has probably caused the unauthenticated assertion, that Magellan had taken with him hogs from the Ladrones. Of this, however, Massimiliano Transilvano, as well as the Breve Narratione di un Portughese, (Ramusio,) are silent; and Herrera, Historia de los Indios, tom. ii. cap. 3., does not mention it. All the authorities agree, that on taking possession of them, there were no four-footed animals. Herrera, l. c., ascribes to these islands, rice, (*y poco arroz,*) evidently without any reason.

east of the Philippines, as far as the one hundred and eightieth degree of longitude. This mild and amiable people are not low in the scale of civilization, and the inhabitants of the Marianas were not inferior to their brethren.

They equalled, at least, in navigation, the most skilful of the Carolinians.* The still existing works of their art of building in Tinian and Saypan, prove that they were in this respect superior to others, and we have discovered something among their antiquities, which seems to indicate the immense advance they had made in civilization before all the other islanders of the Great Ocean. We speak of the invention of money. We have seen ourselves the objects we describe; and explain them on the authority of our friend Don Luis de Torres, the friend of the Indians, and who is perfectly acquainted with their manners.

On a coarse cord of cocoa-bast are stringed pieces of tortoise-shell, of the form of a button,

* We must here mention an error in Dampier's account of the Proas of the Marianas. The boats of the Carolinians sail in reality, only as has been described in Anson's Voyage, and as is already mentioned by Pigafetta, with the out-rigger on the lee-side, and the flat side of the boat under the lee. It was likewise after Anson, that these boats were imitated in England; the rate of twenty-four knots, which Dampier ascribes to them, must be exaggerated, though they are light, quick, and particularly more adapted than ours to sailing close to the wind. We must further observe, which is understood of itself. that the rudder is always under the wind, which is not always observed with respect to the boats of Radack in the drawings accompanying this work.

but as thin as paper, pressed to each other, and extremely polished by rubbing. The whole forms a pliable roll or cylinder about the thickness of a finger, and several feet in length.

These cords are said to have been current as a means of commercial intercourse, and but a very few chiefs had the right to manufacture and issue them.

Plates of tortoise-shell, of the large sea-turtle, are differently pierced in the middle with a large hole, and on the broad thin edge with several smaller holes, or they have only one hole in the middle.

Whoever, probably in swimming, had killed a turtle, (in reality a very hazardous adventure,) brought a plate of the mail to the chief, who, according to the circumstances of the deed and the assistance received in performing it, bored holes in it; the fewer of them the greater was the value. Such trophies then gave the owner a certain right to exchange them, according to established customs, for other property, and passed, in a certain manner, as means of commerce and signs of value.

While the islanders of Guahon, says Crozet, acquired new information by their civilization, they had no improvement to make with respect to the building of their boats, the art of doing which they had inherited from their fathers, and perfectly retained.*

* Nouveau Voyage à la Mer du Sud, par Marion et Duclesmeur, Rédigé sur les Plans et les Journeaux de M. Crozet, p. 204.

Shall we credit this assertion as that of former navigators? It is now quite altered since the time of Anson (1742,) and Duclesmeur (1772.) The present inhabitants no longer know the sea, are no mariners, no swimmers; they have ceased to build boats. They now scarcely hollow out, without skill, the trunks of trees to fish within the breakers. It is the inhabitants of the Carolinas, (Lamureck, Ulea, &c.) who, since the pilot Luito from Lamureck, in 1788, re-discovered Waghal (Guahon) for his islands, come every year, since 1805, with a trading fleet to Guahon, and provide the Spaniards with the requisite boats, which they build on their islands, in exchange for iron. It is also they who, in their own boats, forward the messages from the governor to Tinian and Saypan, and maintain the otherwise difficult communication between the Mariana islands.

There are here at present about ten or twelve of these Carolinian boats, and nobody remembers that similar ones were ever built at Guahon. Have not foreign boats deceived former navigators? Carolinian boats have been cast here in all times, and particularly, in the year 1760—70, a boat from Eap, for so far our accounts, founded on recollection, go back.

The present inhabitants of Guahon have been transformed into Spaniards.* They live and dress

* We expressed a wish to be acquainted with the peculiar manners, plays, dances, of the natives; and the Governor had

like the Tagalese about Manilla, cultivate rice for present use, prepare and drink the cocoa wine, chew the betel, and smoke tobacco, indolently enjoying, to an advanced age , the fruits of the forest, the produce of the fertile earth, and the bounty of heaven.

And how should industry flourish? The Governor of this distant part of the world receives his office for only a short time as a sort of benefice. He has the sole trade of the colony, that is, he retains the considerable sums, in ready money †, which Spain sends for salaries, and for that is bound to give his inferior officers as little and as bad goods as he pleases. ‡ On the other hand, the Indian pays no tribute, cultivates his own tobacco, and has not to enrich the church by tithes.

The galleons from Acapulco now touch but very seldom at Guahon, and, now and then only, the Americans trading on the N. W. coast of America. The present Governor of the Marianas, has a ship of his own, a handsome brig, with which

---

an opera ballet of Montezuma performed before us, in theatrical costumes, which have been, since old times, in the college, in which were the Schools of the Jesuits.

* A robust old man, of eighty-six years four months old, lives in Agana, with his equally aged wife, the only companion of his youth and age; they have about them one hundred and thirty-five descendants, down to the sixth generation.

† About eighteen thousand piasters annually, a sum for which we will, however, not vouch.

‡ Zuñiga, p. 6.

he keeps up an intercourse, and the necessary trade with Manilla, and carries on, besides this, the trade of the *Biches de mer.* He has begun to encourage the Carolinians to bring him this article, which is frequent on their islands, as his pilot, an Englishman, has refused to fetch them on account of the dangerous reefs. This step may have a very great and beneficial influence on the farther civilization of these islanders.

The Jesuits, till the abolishment of their order, remained in possession of the missions which they had founded in the Marianas. They consigned to the flames a part of their books and papers when the Augustines came to succeed them, and then abandoned the field. As there has, for some time, been a want of missionaries, the spiritual care of the Marianas has been given to secular priests. The islands are divided into two parishes; that of Agaña and that of Rota, which latter includes a part of Guahon. Both of them are properly under the bishop of Zebu; who, on account of the too great distance, leaves the administration of them to the archbishop of Manilla.

The parish priests are young Tagalese from Manilla; for whom the Spanish language is sufficient for their duty. They live in Agaña, in the building of the mission.

On the island of Rota there is now a fixed settlement, under the superintendance of an officer, while, on the contrary, there are no habitations on the island of Tinian: it is only visited for the cultivation of rice. We were informed that there were in Tinian, oxen, hogs, and goats; in Saypan, oxen and hogs; and in Agrigan, hogs and goats, in a wild state.

Several Carolinians who have received baptism have settled at Guahon; we found but a few of them there at the time. Some had obtained permission from the Governor to visit their friends in their islands, and had gone there in the preceding year, with the fleet from Lamureck.

It still remains to be explained why natives of the Sandwich islands can be counted among the inhabitants of Guahon on the annexed table.

The reader has found, in another part of this voyage, a circumstantial account of the kidnapping of the people from Easter Island, which was perpetrated by the captain of an American ship, with violence and bloodshed, for the purpose of founding a settlement on the Galepagos islands.

The trade of this ocean makes it desirable for the navigators who possess it, to have similar settlements on the more eastern islands. Their connection with the Sandwich islands renders the stealing of people easy there; and the island of Agrigan, one of the most northern of the Marianas, seemed to be particularly adapted for such a settlement, though it is mountainous, unfit for cultivation, and cannot even feed oxen ; and affords no protected anchoring-place.

Captain Brown, with the ship Derby, from Boston, was in Atooi in the year 1809 or 1810. On this island, he was joined by Mr. Johnson, ship-builder to the king, who had fallen into disgrace, on account of an accident which had happened to a ship. They weighed anchor during the night, and carried off fifteen women who were on board. They approached the island of Oneeheow. A boat brought refreshments from shore. It was expected: seven men who were in it were taken on board, the boat was then hoisted up, and they directed their course to Agrigan. They missed the island; it was to the north: not to lose time in contending against the wind, they attempted to land on a southern island. They did so at Tinian, where they remained in two parties. One party, consisting of Johnson, with four men, and the Sandwich islanders, were to build a boat to sail to Agrigan; the other party, composed of the second mate of the ship, with three men, who had been discharged, intended to convert a long-boat, which they had bought of the captain, into a ship, for the purpose of carrying on commercial speculations on these seas. The Sandwich boat was left behind: both parties went over to Saypan, which island afforded better timber, and there carried on their work. But the Sandwich islanders remembered their liberty, vengeance, and their country. When the mate had finished his vessel, which they intended to make use of to return home, they took advantage, when the party was dispersed and unarmed, to fall upon them; the mate and one white were killed: war raged.

It was, in the mean time, made known in Guahon, that there were strangers in Saypan and Tinian; the Governor, D. Alexandro Parreño, sent thither, and it was in the course of these bloody combats, that, in June, 1810, Johnson, with four whites, two negroes, the seven Sandwich islanders, and the fifteen women, were brought to Guahon, where he himself still remains.

In May, 1815, by command of the captain-general of the Philippines, D. Gose Gardoque, a settlement on Agrigan was broken up, and nearly forty men, of whom one was an American, three Englishmen, and the rest Sandwich islanders, brought to Guahon.

It is well known, from authentic information, that there is already a new settlement on Agrigan. In pursuance of the present order of the captain-general, no obstacle is to be thrown in the way of the settlement; the settlers are only to acknowledge the supremacy of Spain; and a Spaniard is to be sent as chief magistrate. Nobody has, however, yet been sent.

Guahon calls to mind the name of the Governor, D. Thomas, which is known in Europe.

In the Nouveau Voyage à la Mer du Sud he

is mentioned with much praise; and the Abbé Raynal, in his manner, consecrates him to immortality. La Peyrouse, soon after, found him in the hands of the inquisition, at Manilla and attributes it to the praises of the philosopher. We, however, doubt, from better local knowledge, that the blame of this injustice can be entirely attributed to the French author.

The inquisition, like chance, falls among the rich and high, upon every one against whom information is brought; and it is customary for the women, in case of domestic dissensions, to arm the holy tribunal in their cause. The property of the person condemned falls to the tribunal; and only the poor and obscure man enjoys security.

EXTRACT

FROM THE ARCHIVES OF SAN YGNACIO DE AGAÑA.

In the year		Number of Inhabitants.	Increase.	Decrease.
1783	Of the first of the Government of D. Felipe de Cerain.	3231	—	—
1784	... ...	3213	—	18
1785	... ...	3292	72	—
1786	... ...	3301	9	—
1787	The first of the Government of D. Jose de Arlequi y Leoz.	3344	43	—
1788	... ...	3433	89	—
1789	... ...	3501	68	—
1790	... ...	3564	63	—
1791	... ...	3630	66	—
1792	... ...	3680	50	—
1793	... ...	3584	—	96
1795	The first of the Government of D. Manuel Muro.	3500	—	84
1796	... ...	3643	143	—
1797	... ...	3789	146	—
1798	... ...	3935	146	—
1799	... ...	4001	66	—
1800	... ...	4158	157	—
1801	... ...	4245	87	—
1802	... ...	4249	4	—
1803	The first of the Government of D. Vincente Blanco.	4303	54	—
1804	... ...	4308	5	—
1805	... ...	4354	46	—
1806	The first of the Government of D. Alexandro Parreño.	4442	88	—
1807	... ...	4545	103	—
1808	... ...	4690	145	—
1809	... ...	4804	114	—
1810	... ...	4845	41	—
1811	... ...	4958	113	—
1812	The first of D. José de Medinilla y Pineda.	4921	—	37
1813	... ...	5049	128	—
1814	... ...	5232	183	—
1815	... ...	5315	83	—
1816	... ...	5389	74	—
	Increase		2393	234
	Decrease		234	
	Net Increase ...		2158	

San Ygn. de Agaña, capital of the Mariana islands, the 27th of Nov. 1817.

TABLE OF THE VILLAGES, HOUSES, AND INHABITANTS ON THE MARIANAS,

Drawn up by Don José de Medinilla y Pineda, Lieutenant of the Royal Regiment of Infantry, Justicia Mayor, Civil and Military Lieutenant-Governor and Captain-General of those Islands and their jurisdiction, in this Year 1816.
Accompanied with Observations on the Increase and Decrease since the last Year, and the different Classes.

The Island of Guahon.	Houses.	Civil Officers and Military.	Spaniards and Mestizos.		Natives of the Philippines and their descendants.		Indians.		Mulattos.		Indians from the Sandwich Islands and the Carolinas.		Total in the Year 1816.	Total in the Year 1815.	Increase.	Decrease.
			Men.	Women.	Men.	Women.	Men.	Women.	Men.	Women.	Men.	Women.				
CAPITAL.																
San Ygnacio de Agana.																
ITS FOUR QUARTERS.																
Santa Cruz, San Ygnacio, San Nicholas, San Roman.	448	147	535	568	670	764	188	172	10	9	23	29	3115	3062	53	—
DEPENDENCIES.																
Anigua	44	—	—	1	1	2	116	118	—	—	—	—	238	233	5	—
Asan	28	—	—	—	—	4	64	48	—	—	—	—	116	112	4	—
Tepungan	14	—	—	—	—	1	36	34	—	—	—	—	71	67	4	—
Mungmung	15	—	—	—	1	3	44	36	—	—	—	—	84	90	—	6
Sina Sana	36	—	—	1	—	2	99	82	—	—	—	—	148	188	—	4
SEPARATE VILLAGES.																
Agat	45	—	—	—	4	5	118	112	—	2	—	—	241	244	—	3
Villa de Umata	34	—	—	1	6	5	86	84	4	3	—	—	189	184	5	—
Merizo	52	—	—	—	7	3	144	138	—	—	—	—	292	288	4	—
Unarasan	43	—	—	1	—	—	99	104	—	—	—	—	204	201	3	—
Pago	40	—	—	2	—	2	98	88	4	6	—	—	200	204	—	4
ISLANDS.																
Rota, and Tinian.	103	—	—	—	2	2	228	223	—	—	—	—	455	442	13	—
Total	902	147	535	574	691	793	1320	1239	18	20	23	29	5389	5315	91	17

The census of both sexes, and all classes, which was made with the greatest accuracy and minuteness, gave the number of souls, 5389, being an increase of 74 since the first of February, 1816. The number of houses was 902. (Signed) José de Medinilla y Pineda, *Justo de la Cruz*.
San Ygnacio de Agana, Mariana Islands 4th of March, 1817.

RESPECTING OUR KNOWLEDGE OF THE FIRST PROVINCE OF THE GREAT OCEAN.

New Sources Kadu, Don Luis de Torres Geographical View, (with a chart,)

According to the last discoveries of Saavedra, in 1528; Villalobas, in 1542; Legaspi, in 1565, and others. After the discovery of the Carolinas (perhaps Eap) by Lazeano, in 1686, Paul Clain, the Jesuit, collected, in 1697, in the Philippine islands, the first certain information respecting the islands, which were afterwards called the Carolinas, from natives of these islands, who were driven by storms to Samal. We also learn that those islanders frequently visited these coasts, sometimes by accident and sometimes by design.

Lettre du P. Paul Clain, Lettres Edifiantes, tom. i. p. 112. Aux Jesuites de France, Charles Gobien, tom. vi. with a chart, by Serrano, which deserves no attention.

The zeal of the missions is awakened; all the monarchs of Europe are summoned to contribute to spread the doctrines of Christianity. Several ships are fitted out at Manilla, which kind fortune, favouring the people, and to preserve their happiness and independence, diverts from their object. At last, Fathers Cortil and Duperon land in Sonsorrel in 1710: the winds and currents soon carry the ship out of sight; the missionaries are abandoned; and every further attempt to come to their assistance is ineffectual.

Aux Jesuites de France, J. B. du Balde, tom. vi. Relation en Forme de Journal, tom. vi. p. 75. Lettre du P. Cazier, tom. xvi.

Father Jean Antoine Cantova collected, at Guahon, in 1722, from the islanders of Ulea and Lamureck wrecked there, a circumstantial account of the Carolinas, and gives a chart of these islands, which deserves the greatest regard; his heart was inflamed with a desire to spread the Gospel there.

Lettre du Père J. A. Cantova, tom. xviii. p. 188. with a chart.

The historians of Manilla have carefully collected these histories from the sources.

Historia de la Provincia de Philipinas de la Compania de Jesus, par el P. Pedro Murillo Velarde, Manilla, 1749, tom. ii. Historia General de Philipinas par Fr. Juan de la Concepcion, tom. ix. c. 4. p. 151., and tom. x. c. 9. p. 239.

We have borrowed the following from the latter.

Cantova succeeds in being sent to the inhabitants of the Carolinas. In 1731, he is brought to Mogemug, with Father Victor Uvaldec, from Guahon, and a mission is founded on the island of Falalep. Father Victor makes a voyage to the Marianas; when he returns, in 1733, with new assistance for the mission, he finds the place where it stood ravaged and desolate. He now continues his tedious route to Manilla. "They learnt from a prisoner whom they took, that ten days after the departure of Father Victor, on the 9th of July 1731, Father Cantova was called upon, on pretence of baptizing a grown-up person. He went thither with two soldiers, and found all in arms. They alleged that he wanted to establish a new law against the old one, and against their customs; they pierced him with three lances, two in the sides, and one in the heart; they likewise killed the two soldiers, and threw them into the sea. They stript the Father, much astonished that he was so white, and hid him under a small roof.* They afterwards fell unexpectedly on those who had remained at Falalep, who were only able to discharge their small cannons; killed four Indians, and wounded others with the sword: but their defence was in vain. All the Spaniards on the island, who were fourteen in number, were massacred; and only a young Tagalese, the sacrist of the Father, and whom the chief of the island had adopted for a son, was spared.

"This same prisoner informed them, that the confidant of the Father, one of the name of Digal, whom he had baptized at Guahon, was the principal leader of this revolt."

* So they bury their own dead. The Father was treated as a prince, the soldiers as common people.

Thus ends the history of the missions in the Carolinas.

We have since become acquainted with a single group of these islands, by An Account of the Pelew Islands, from the Journals and Communications of Captain Henry Wilson, by George Keate, Esq.

Burney, in the first chapter of the fifth volume of his Chronological History of Voyages, circumstantially relates, from the original authorities, what relates to the Carolinas. At the death of Cantova, he mentions a memorial of the Governor of the Philippines, but of which we could not obtain a sight. The fifth chapter contains a complete representation of our geographical knowledge of the islands, which the Spaniards comprehend under the name of *las Carolinas.*

We are induced to unite in one point of view, and under the name of the Western, or First Province of the Great Ocean, the Carolinas, to which the more westerly groups are to be added, with the islands lying more to the east, nearly under the same latitude, as far as those which Krusenstern calls after their principal discoverers, Gilbert's and Marshal's islands, together with the Marianas, to the north of the Carolinas.

Krusenstern, in his Contributions to Hydrography, Leipzig, 1819, has collected the discoveries made by later navigators in these seas under different heads, in pages 94 to 121, and has treated them with much learning. He has made great use of the Memorias por Don Joseph Espinosa y Tello. Madrid, 1809.

Tuckey, (Maritime Geography and Statistics, London, 1815,) by neglecting to point out the sources from which he derived his determination of doubtful points (the islands of Lamurca, Hogolen) has deprived his work of all authority.

Arrowsmith's Chart of the Pacific Ocean, with the additions, to 1817, seems to us to be of greater authority.

As we are now going to communicate the information respecting the islanders and people of this province in particular, derived from our own experience, and the information we have collected, this seems to be the place to give an account of the new authorities which we have to adduce.

These are the communications of our friend and companion Kadu, and those of D. Luis de Torres of Guahon, which are a supplement to Cantova's letter and chart.

It was at the beginning of the year 1817, in the extreme east of this province, in the group of Otdia and Kawen, on the island-chain of Radack, that we formed an acquaintance, and confirmed our friendship, with the amiable people who inhabit it. When we were afterwards sailing to the group of Aur, of the same chain of islands, the natives came in their boats to meet us. As soon as we had cast anchor, and they came on board, a man stept out from among them, who was distinguished from the others in many respects. He was not regularly tattooed like the Radackers, but wore indistinct figures of fish and birds, singly and in rows, round the knee, on the arms and on the shoulders. He was of a more compact make, and of a lighter colour, and had more curly hair than they. He addressed us in a language that was quite different from that of Radack, and sounded entirely foreign to us, and we were equally unsuccessful in making him understand the language of the Sandwich islands. He made us comprehend that he intended to remain in our ship, and to accompany us in all our future voyages. His request was readily complied with. From that hour he remained on board our ship, and only once, at Aur, went on shore, with permission, and remained with us, a faithful companion, treated like the officers, beloved by every one, till our return to Radack, when, suddenly changing his intention, he resolved to settle there, to become an inhabitant, and to be a distributor of our gifts to our poor friends. Nobody could be more thoroughly sensible of the humane object of our mission than he was.

Kadu, a native of the island-group of Ulea, to the south of Guahon, not of noble birth, but a confidant of his king, Toua, who employed him to carry his commissions to the other islands, had on former voyages become acquainted with the chain of islands with which Ulea trades, from the Pelew islands in the west to Setoan in the east. He was on his last voyage from Ulea to Feis, with two of his countrymen and a chief from Eap, who was returning to his native place, when storms drove the boat from her course. The mariners, if we may credit their very uncertain reckoning, drifted about in the open sea for eight months. Their scanty stock of provisions lasted them three months; for five months they lived without fresh water, merely on the fish they caught. To alleviate their thirst, Kadu dived into the depth of the ocean, and brought up in a cocoa shell cooler water, which, according to their opinion, was likewise less salt. The north-east monsoon at length blew them on the group of Aur, of the Radack chain, where they fancied themselves west of Ulea. Kadu had received information from an old man in Eap of Radack and Ralick; some mariners from Eap are said to have been once cast upon Radack on the group of Aur, from whence they found their way back, by way of Nugor and Ulea, to Eap. The names of Radack and Ralick were also known to a native of Lamureck, whom we met with at Guahon. Boats from Ulea and the surrounding islands are frequently cast upon the eastern island chains, and there are still living on the group of Arno, of the Radack chain, five natives of Lamureck, whom a similiar fate brought there in the same manner.

The chiefs of Radack protected the strangers against the rapacity of their people, whose avarice was excited by the iron which they possessed.

The more noble sentiments are always found among the chiefs.

The inhabitants of Ulea, who live in greater prosperity and have a more extensive trade than the Radackers, are in many respects superior to them. Kadu enjoyed a certain consideration at Radack. When we visited these islands, he might have arrived there about four years before. He had two wives at Aur, and had a daughter by one of them, who was just beginning to speak.

Our appearance caused terror and consternation at Aur, where no information had yet been received respecting us. The experienced Kadu, who was at that time on a distant island of the group, was immediately sent for, and they desired his advice how they should treat the mighty strangers, whom they were inclined to consider as wicked cannibals.

Kadu had learnt much of the Europeans without ever having seen one of their ships. He encouraged his friends, warned them against theft, and accompanied them to our ship with the firm resolution to remain with us, hoping through us to see his dear native country again, as an European ship had once been at Ulea at a time when he was absent.

One of his countrymen, and companion in misfortune, who was with him, tried in vain to dissuade him from his purpose, and his other friends, equally in vain, assailed him with anxious representations: he was immoveable. Another companion of Kadu, the chief of Eap, whom we met with in the suite of king Lamary, at Udirick, conceived the same resolution, the same hope as our friend. He was a weak venerable old man; his request was not attended to. It was difficult to persuade him to leave the ship, where he persisted in remaining in tears, in the composed attitude by which he meant to make us sensible of his resolution. We represented to him his age, and the fatigues of our voyage; he remained inflexible. We then told him that our stock was taken in only for a certain number of people: he proposed to us to leave our friend Kadu here, and take him in his stead.

We cannot but commend the easy and becoming manner in which Kadu conformed to our customs; it was difficult for him to understand the new situation in which he was placed. He, a man of low rank, was suddenly placed among strangers so superior in power and wealth, treated like one of their officers, waited upon by the sailors in the same manner as the captain. We will not conceal the mistakes into which he sometimes fell, but which he so quickly and easily corrected, that they merited no severe reproof. When, shortly after he was received among us, chiefs from Radack came on board, he treated them with haughtiness, and assumed a behaviour which became only them. Some innocent raillery on their part was no more than he deserved: it never occurred a second time. He, at first, tried to imitate the walk and the manners of the captain, but gave it up of himself. It is not remarkable that he should, at first, consider the sailors as slaves. He once ordered the waiter to bring him a glass of water; the latter took him by the arm, led him to the water-butt, and gave him the cup out of which the others drank. He reflected, and studied our relations, and the spirit of our manners, to which he soon learnt to conform, and to adopt our behaviour at table, as well as in general.

Kadu learnt only by degrees the power of our spirituous liquors; some of us imagined that he, at the beginning, got the sailors to give him brandy. Some time after, when a sailor was punished, he was told that it was because he had privately taken some of the fire, (the name which he gave to brandy.) He never after drank brandy; and wine, of which he was very fond, but with great moderation. The sight of drunken men at Oonalashka made him carefully keep a guard upon himself.

At the beginning he adjured the winds in our favour, according to the custom of Eap; we laughed, and he soon laughed himself at these adjurations, which he afterwards only repeated in joke, to amuse us.

Kadu had feeling, sense, and wit; the more we became acquainted with him, the more partial we were to him. We found in his amiable character only a certain indolence to contend with, which counteracted our views; he only liked either to sing or sleep.

When we tried to draw from him information

about the islands which he had visited, or respecting which he had some knowledge, he merely answered the questions which we put to him; and unwillingly the same question a second time, referring to what he had previously said. When, in the course of conversation, new particulars were mentioned, which we reproached him for having concealed, he coolly answered, "You did not ask me that before;" and besides this, his memory was not correct. His recollections revived by degrees, as occasion called them forth, and it appeared to us also, that the multitude and diversity of the objects which drew his attention effaced earlier impressions. The songs, in different languages, which he sung, and which he learnt from the people among whom he had resided, served him, as it were, as a book, in which he sought explanation or confirmation of his assertions.

Kadu kept his journal by moons, for which he made a knot in a string; this journal appeared to us to be very irregularly kept, and we could not understand his reckoning. He was not unapt at learning, nor without curiosity. He seemed to comprehend what we endeavoured to explain to him, respecting the figure of the earth, and the art of navigation; but he had no perseverance, was soon tired, and went back to his songs. He took some trouble to learn writing, the secret of which he comprehended, but was not able to succeed in this difficult task. What we said to encourage him, perhaps, had a contrary effect. He interrupted his attempts, took them up again, and, at last, renounced them entirely.

He appeared readily to understand what we told him respecting the social institutions of Europe; of our manners, customs, and arts; but what most struck him, was the peaceable chivalrous nature of our voyage, with which he connected an intention to teach the new-discovered people what might be good and useful to them. By this, it is true, he chiefly understood what served for food; but he was also sensible that our superiority depended upon our greater knowledge in general; and he assisted, to the utmost of his power, our researches, where they would have appeared very idle, even to many of the better informed among ourselves.

When we arrived at Oonalashka, and he had contemplated this sterile country, entirely destitute of trees, he hastened to request us to plant some cocoa-nuts, which we had still on board, and to which he offered to add some belonging to himself, in suitable situations; he urged us to make the trial, representing the misery of the inhabitants, and was with difficulty persuaded that it would be quite useless. Nature, above all, attracted his attention and curiosity. The oxen in Oonalashka, which, for the first time, put him in mind that he had seen them before on the Pelew islands, constantly employed him, and he went every day into the fields to look at them. On the whole voyage nothing gave him more pleasure than the sight of the sea-lions and sea-bears on St. George's island. *

As Kadu, during the voyage, never neglected carefully to collect pieces of iron, broken glass, and every thing overlooked by us, which might be valuable to his countrymen, he looked on the shore at Oonalashka chiefly for stones, which might serve for whet-stones. We only once saw this mild man angry; it was when, in the course of our voyage, he looked to no purpose for these stones in the place in the ship where he had put them, and his complaint met with little attention; his sense of justice was wounded.

Kadu, in his poverty, was generous and grateful. He served such of us as made him presents; and made use of the opportunity at Owhyee, by the barter which he judiciously made with the little articles which we had given him, to make presents to us and the sailors who had obliged him, such as might be agreeable to each. He retained nothing for himself, except that with which he intended one day to enrich or to please his countrymen: thus he had left every thing he possessed to his friends at Radack; all but one treasure, a necklace, which

* When, after returning on board from the island of St. George, we conversed about the sea-lions, in humourously imitating the voice and gait of which Kada amused himself and us, he was asked, with apparent seriousness, whether he had looked at the nests and eggs, under the rocks on the sea shore. However little versed he might be in the Natural History of Mammalia, the question surprised him, and he soon discovered that it was a joke, which made him laugh heartily.

he wore for a long time among us. One day, smiling, with a tear in his eyes, he entrusted us with the secret of this necklace: he fought at Tabual (an island in the group Aur of the Radack chain) in the ranks of his friends against the enemy, who came from Meduro and Arno; there he gained the advantage over his opponent, and was about to pierce him as he lay at his feet, when his daughter rushed forward and seized his arm; she obtained of him her father's life. This girl promised him her love; Kadu privately brought her considerable presents to her island; and he wore, for her sake, this pledge of love which she had given him on the field of battle.

We must particularly mention two traits in Kadu's character, his utter aversion to war, murder, and his delicate modesty, which did him honour, and which he never violated while among us.

Kadu abhorred bloodshed, though he was no coward. He had on his breast scars of the wounds which he had received in the war in defence of Radack; and as we were preparing our arms for landing on the island of St. Lawrence, and he was informed that this was not out of an hostile intention, but for self-defence, in case of danger among a people with whose sentiments we were not acquainted, and with whom we merely wished to deal to our mutual advantage, he demanded a sword, with which he might defend us if attacked, as he had not sufficiently practised firing with the musket while at Oonalashka.

With respect to the other sex, Kadu maintained a reserve worthy of praise. He remained at a distance from women who had other husbands; he, in fact, possessed great notions of propriety. What he learnt on Owhyee disgusted him, and he spoke freely of the want of good morals in the Pelew islands. When drawn into the free conversation of men, he always took part in such a manner as to keep within the strict bounds of decorum.

The most lively sense, and the greatest talent for wit, are found among those people who are the least remote from nature, where the mildness of the climate affords to man an easy and pleasant life. Kadu was particularly witty, but knew how to observe due limits in harmless pleasantry, and ingeniously contrived to conciliate by little services and presents those whom his raillery had offended.

Our friend often repeated in the course of our voyage, that he intended to remain with us to its destination, and even should we discover his dear Ulea, he would not quit us, but accompany us to Europe, from which we might promise him a passage home to Ulea, as the purposes of commerce regularly lead our ships to the Pelew islands, whither the boats of Ulea regularly trade. We ourselves were still unacquainted with the other way by Guahon, but he cherished the wish, and this might have been fulfilled at Guahon, to find an opportunity in one of the islands known to him of sending an account to Eap of the fate and the present residence of the chief of that island, his companion in misfortune at Radack; his idea was, that his countrymen should build a ship, and come there to fetch him. He seriously occupied himself with these ideas.

We endeavoured to collect useful animals and plants, layers, and seeds of various kinds, which we wished to introduce into Radack. Kadu knew very well that we intended to call there, and remained firm to his determination; we advised him to inform himself of every thing that could be useful at Radack, that he might instruct our friends, and teach them what advantages they might derive from our gifts, and how to manage them. He entered into our plan it is true, but the object was too remote, and thoughtlessness and indolence caused him to reap in this abode of pleasure but little advantage, which he afterwards blamed himself for having neglected.*

We arrived at Radack, and landed at Otdia amidst the shouts of our few friends who had not engaged in the war. Kadu was unwearied, and assiduously assisted us in planting, sowing, and managing the animals, and in explaining and instructing the natives in every thing necessary. He was then still firmly resolved to remain with us.

When every thing was ready at Otdia, Kadu went to Ormed, the island of the old chief Laergass, to plant there likewise a garden. In this ex-

* Kadu soon learnt to converse with the people of Owhyee; and he himself made us remark the great similitude of several words in their language, with the language of the islands in the first province of the Great Ocean.

cursion, which was undertaken in boats belonging to people of Radack, he was accompanied only by the writer of this account. At Ormed the day was spent in labour, and the evening in social conversation. The women sung to us the many songs which had been composed on us during our absence, and in which our names were commemorated. Kadu told them of his travels, and blended lively fables with his narration. He distributed among them the presents which he had collected in the course of his voyage for his friends. On the following day, the last of our stay at Radack, immediately when the boat which carried us back to the ship was under sail, Kadu, whose lively serenity was changed into gloom, declared that he was determined to remain at Otdia, and would go no farther with the Rurick. He expressly commissioned his friend to acquaint the captain with this new unalterable resolution, and declining all representations to the contrary, explained the reasons which actuated him. He said he would remain at Otdia to be the guardian of the animals and plants, which, without him, would be neglected out of ignorance, or be destroyed without any use to these unthinking people. He would bring it about, that our gifts might procure the needy inhabitants of Radack a sufficiency of food, that they might no more be obliged to kill their children, and renounce that custom. He wished to bring about, that peace might be established between the northern and southern groups of Radack, that men might not murder one another. He intended, when the animals and plants had sufficiently multiplied, to build a ship and go over to Ralick, and to communicate to that country the benefit of our gifts. He would ask from the captain, returning every thing he had received from him, only a spade to dig the ground, and some useful tools. He depended in his undertaking on the assistance of his countryman and companion in misfortune, for whom he would send to Aur, where he then resided. He was also to bring his daughter with him, who, as he now learnt, was very melancholy since his absence, and would take no rest. His wives had taken other husbands, only his child was an object of his tenderest affection.

Kadu now regretted the many useful things which he had neglected to learn at Owhyee, and asked our advice in these last moments on several subjects, to which he paid the greatest attention.

The boat in which we performed this trip against the wind, was a bad sailer; the sun was already sinking to the horizon when we came up to the ship, and happily found the captain on board. When Kadu's resolution was made known, he saw himself instantly and unexpectedly in possession of immense riches, such as are the objects of desire to princes and people in this part of the world, κμητον δε σιδηρον. The affection was made manifest which he enjoyed among us, and every one was seen busy in adding out of his own stock, to the heaps of iron, tools, and other useful things which had been collected for him. Specimens of mats and stuffs from Owhyee; patterns of straw-hats, &c., were not forgotten.

While Kadu was engaged in packing up his bed-clothes and linen, he carefully separated his winter clothes, and offered them as a present to the sailor who had attended him, which the latter however refused.

The sun had already set, when Kadu was brought on shore with his treasures. Time would not permit us to give him a written testimonial. Only an inscription on a copper-plate, nailed to a cocoa-tree at Otdia, records the name of the ship and the date.

Kadu was then installed before the assembled inhabitants of Otdia, as *our* man to whom our animals and our plantations were confided, and who besides was charged with our presents to Lamari. It was promised, that we, who had been already three times at Radack, should return after some time, to look after him, and to demand an account. As a confirmation of our promise, and as a sign of our power, (for we had hitherto only given signs of mildness and friendship,) we fired off two cannons and a sky-rocket when we returned late in the evening to our ship.

When we weighed anchor next morning, our friend and companion was employed with the animals on the shore, and frequently cast his eyes towards the departing vessel.

One of the songs which Kadu frequently sung

among us, celebrated, in the language of Ulea, the names of Samuel Borman (he pronounced it Moremal) and Luis. This song referred to the European ship which visited Ulea at the time when Kadu was absent on his voyage. Waghal appeared in Kadu's account a large country, where there were oxen, iron, and other riches in abundance, whither King Toua once made a voyage, and from whence he brought three cannon-balls, (two-pounders.)

As soon as we landed at Guahon, we recognized the Waghal in this island, and the Luis of that song came friendly to meet us in the person of Don Luis de Torres, from whom we copy the following account, while we recollect him with sincere love and gratitude.

Luito *, a mariner of the islands to the south of Guahon, whose name is still cherished among his countrymen, found again, in the year 1788, in two boats, the way to Waghal or Guahon, of which a song appears to have preserved the remembrance from ancient times. Encouraged by the success of his first voyage, and the kind reception he met with, he came again, in the year 1789, with four boats, requesting the Governor to allow him to come there annually. The four navigators, when they had prepared themselves for their return, disagreed about the rout they should steer; they se-

* Compare Espinosa and Krusenstern's Contributions to Hydrography, page 92.

parated, and none of them ever returned to his native land.

After this, the intercourse that had been begun was interrupted.

In the summer of the year 1804, the ship Maria, from Boston, under Captain Samuel William Boll, supercargo Thomas Borman, went from Guahon, for the discovery of the Trepang in the Carolina islands. Don Luis de Torres went as passenger on board the Maria, in the hope of seeing the islanders again, to whom he was attached, to do them good; to learn from them why they had discontinued their visits to Guahon, and to prevail on them to return.

On this voyage, the following points were geographically determined, according to the journal of Don Luis:

A shoal of twenty-four fathoms in 8° 20′ north latitude, and 149° east longitude from Greenwich.

The desert island of Piguelao (D. L. de T.), Bigellé (K.), in 8° 6′ north latitude, and 147° 17′ east longitude, (wanting in Cantova's charts).

The shoal Oraitilipu of twelve fathoms, in the same latitude, half-way to the desert island of Fallao (D. L. de T.), Fahueu (Cantova).

Fayo (K.), in 8° 5′ north latitude, and 146° 45′ east longitude.

The small low group Farruelap (D. L. de T.), Faroilep (Cantova), Fatoilep (K.), in 8° 3′ north latitude, 144° 30′ east longitude; and, lastly,

The group of Guliai (D. L. de T.), Ulee (Cantova), Ulea (K.), *Olä* (according to the pronunciation of Radack,) in 7° north latitude, and 144° east longitude, into which group the Maria penetrated, and where she passed some time.

Don Luis de Torres took the opportunity when he was at Ulea, the language of which he understood, and whose amiable inhabitants he highly valued, to obtain thorough, and authentic information of the most able of the people, concerning themselves, and the tribes related to them, with whom they traded. He has drawn a chart of Ulea, according to the most experienced navigator among the natives, attending to the courses which they sail, with all the islands that were known to them, on it: the coincidence of which, with the chart of Cantova, which was unknown to him, is very remarkable. He has since lived at Guahon, in constant communication with his friends there, and yearly seen the skilful navigators who conduct the trading fleet from Lamureck to Guahon. We regret having had so little time to profit from his experience and information, which he so liberally opened to us, and we expect a supplement far richer than our harvest from the French expedition under Captain Freycinet, who has promised a longer stay at Guahon, and with the learned gentlemen attending whose expedition we have conversed upon the subject at the Cape.

Don Luis de Torres learnt at Ulea that the not returning of Luito, in the year 1789, had been laid to the charge of the Spaniards in Guahon. The islanders, on being better informed, promised to re-establish the interrupted trade; and they kept their word.

An Englishman, who was a passenger on board the Maria, and was called by Don Luis, Juan, settled at Ulea. Kadu had known him there, by the name of Lisol; he had taken a wife, by whom he had one child. According to his account, this Lisol, at a later time, when Kadu was absent, was fetched away again by some ship; but from the information of Don Luis respecting him, he died at Ulea.

Don Luis de Torres, in this voyage, had sought to introduce oxen and hogs, and several useful plants at Ulea. The natives in the sequel purposely extirpated the oxen and hogs, because they were not only useless, but hurtful to them. The oxen bit off the young cocoa-trees, and the hogs damaged the taro plantations. Of the plants, only the ananas had thriven; as it bore fruits, and the people were much rejoiced at it, and as every one wanted to possess the plant, it was transplanted so frequently that it at last died.

Since the voyage of Don Luis, no new misfortune has interrupted the communication. The Carolinians coming to Guahon become yearly more numerous. Their fleet of boats from Ulea and the surrounding groups, consisting of Lamureck and Setoan, collects at Lamureck. The voyage is undertaken from thence in the month of April;

the distance to Fayo, the desert island, where they stop for some days, is reckoned to be two day's voyage, and from Fayo to Guahon three days. They return likewise by way of Fayo and Lamureck. Their time to return is in May, at the latest in June, before the west monsoon, of which they are much afraid, sets in.

Kadu mentioned an undertaking of the chief of Fatoilep, to sail directly from this group to Waghal (Guahon). He wandered long about the sea, and arrived at last at Moge-Mug, without having found that island, and thence he returned home.

The fleet once missed Guahon, and was driven under the lee of that island. The seamen discovered their error in time; and, after contending with the wind, reached their destination, only with some delay.

This long voyage was once undertaken in a very small boat, which carried only three men. It sailed better than the two large vessels that accompanied it. The seaman, Olopol, of Setoan, brought it as a present to Don Luis. Olopol died at Agaña, and we ourselves have seen the boat.

Toua*, the king of Ulea, came himself to Guahon, in the year 1807.

It was also in this year, or in the following, that a boat, from the eastern island of Tuch was

* Don Luis de Torres calls him Roua, as he calls the island Rug, which we, according to Kadu, write Tuch.

driven to Guahon. There were fifteen men on board; the pilot's name was Kulingan. The strangers were well received; but a procession, which took place during their stay, accompanied with salutes of artillery, spread fear and terror among them. They hid themselves in the wood; and in the same night, without any provisions, went again to sea. Fortunately, they met, in their flight, the fleet coming from Lamureck, which supplied them with provisions, and gave them the necessary directions for their voyage home.

The fleet, in 1814, consisted of 18 sail.

The Carolinians procure, at Guahon, iron, glass-beads, cloths, &c., in exchange for boats, shells *, and curiosities: the trepang may become an important branch of commerce. They are received in the most hospitable manner by the natives, during their stay at Guahon.

Don Luis de Torres has undertaken, with pleasure, to inform Kadu's friends at Ulea of his misfortunes, and his present abode; and to send them our presents in his name.

Don Luis de Torres has given us farther information of a large and high island, of an unknown name, which was seen by the Brigantine San Antonio de Manilla, Captain Manuel Dublon, on her voyage from Manilla to Guahon, on the 10th of

* These shells, among which are some of the most beautiful kind, are sent by the Governor of Guahon to Manilla; from which place they are procured for our museums and collections.

December, 1814, in 7° 20′ north latitude, 151° 55′ east longitude. There is a very high mountain on it.

We had heard Kadu sing a song of Feis, which related to a ship, with which the islanders had traded in sight of their island, though she did not stop there. It celebrated the names of Jose Maria, and Salvador. We learned at Guahon, that in the year 1808 or 1809, the Modesto of Manilla, Captain Jose Maria Fernandez, which looked for the Pelew islands, to collect trepang, had missed them, and come in sight of Feis. When the Modesto afterwards reached the Pelew islands, she met there one of the natives of Feis, with whom she had traded while at sea; to continue the commerce he had hastened on before the ship. Don Jose de Medinilla y Pinedo, governor of the Mariana islands, was on board the Modesto. We tried, at Manilla, but in vain, to obtain further particulars respecting this voyage.

We relate here an event, after our friend Kadu, which may excite some interest. Six white men, who wore clothes, once landed at Eap, in a small boat fastened together with wooden pegs, without any iron. This boat was, in other respects, made entirely in the European manner. The strangers were hospitably received. One of them, named Boëlé, was adopted as a son, by Laman, the chief of the territory of Kattepar. Boëlé remained on the island, when the five others, after a stay of a few months, went again to sea. Kadu, who shortly after came to Eap, became acquainted with this Boëlé. He went naked on the island, and was tattooed on the thighs.

The island-chain of Radack will first occupy our attention. With what we know from our own experience, we shall illustrate Kadu's account, in whose accuracy we were confirmed in the last visit we paid our friends. With Radack, are naturally connected,

1. The island-chain of Ralick, which is situated a little to the west, and is perfectly well known to the Radackers;

2. The islands of Repith Urur;

3. Bogha, of which they were informed by people drawn out of their course; and

4. The islands discovered by the Cornwallis frigate, in the year 1809; which Arrowsmith is inclined to consider as the Gasparrico of ancient charts. A desert group, lying north of Radack, which we found again.

The island-chains of Radack and Ralick lie in the part of the sea occupied by Marshall's islands, (Lord Mulgrave's range, and the adjoining islands.)

Captain Marshall in the Scarborough, and Captain Gilbert in the Charlotte, saw these same Islands, in 1788. The former, whom Krusenstern follows, gives them (Voyage of Governor Phillip, London, 1790, p. 218, &c.) a more western situation than the latter, whose original charts and journals Arrowsmith possesses, and follows. No scientific geographical work can be undertaken on the islands of these seas, without consulting these documents. On account of the differing determinations of the two captains, and the different names which each applies to his islands, it is a difficult task to compare their accounts with each other, and with the discoveries of other navigators; and we leave it to more able geographers. They must decide which of these islands, that are here marked by the name given them by the natives, (these are permanent,) were formerly known to our navigators; and which of the islands seen by them, though in the neighbourhood of Radack, have, notwithstanding, remained unknown to the Radackers. The navigator who is content to name, at his own discretion, the islands he discovers, and whose situation he determines, inscribes his name on the sand. He who learns and preserves the true names of his discoveries, ensures the stability of his work, and really assists to erect the edifice, while the others only furnish the stones.

We did not find among the people of Radack any knowledge of Gilbert's islands; that is, of the islands south of Radack; unless which, for many reasons, (the course of the winds, &c.) appears to us doubtful, we should place Repith Urur there.

From Marshall's reports of the south and north chain of the islands discovered by him, they appear to us to be similar in all respects, and to be inhabited by the same people, only that the southern islands are more fertile and populous than the northern, as we ourselves found it to be the case at Radack; and have every reason to believe it to be the same in all the archipelagos of these seas.

Los Pintados and *Los Buenos Jardines* of Alvaro de Saavedra, 1529, are under the latitude 7° to 8° or 10° north, seeming to lie far to the east of Radack. The description of these islands, which have vanished from our charts, and that of their inhabitants, call upon us to mention them here.

We have ourselves observed nature at Radack, and have lived with its inhabitants. Being conversant with this nature, and with the people, the intelligence which we have to give about the Carolina islands will appear more clear.

The Carolina islands will be the subject of another chapter. We shall endeavour, with the aid of our friends Kadu, and Don Luis de Torres, to overlook from Ulea, all the surrounding islands, and attend an amiable people, versed only in the arts of peace, on their courageous voyages. We shall carefully compare our information with that of the Jesuits, and particularly with the estimable reports of Cantova.

We only enumerate these islands, and make use of the geographical remarks which offer themselves to us. This part of our work, like the chart of Tupaya, and the information collected by Quiros from the natives of Taumaco and other islands, may contain hints not quite unworthy of the notice of future navigators.

The charts of Cantova and Don Luis de Torres which we have added, will serve to illustrate our information. The modern discoveries which we have quoted are to be referred to in the original, or in the above mentioned hydrographical works, and particularly on the charts of Arrowsmith and Krusenstern.

Ulea (K.), *Olä*, according to the pronounciation of Radack, Ulee (C.), Guliai (T.), and according to him in 7° north latitude, and 144° east longitude. The thirteen islands mentioned by Wilson, in the Duff, in 1797, 7° 16′ north latitude, and 144° 30′ east longitude. (?)

A principal group of low islands. The names of eleven islands are marked on Cantova's original chart; Kadu has named four-and-twenty to us, and passed over the smaller uninhabited ones. Namely,

According to Kadu.	*According to Cantova.*
Ulea,	Ulee.
Raur,	Raur.
Pelliau,	Peliao.
Marion,	Mariaon.
Thageiliip,	Tajaulep.
Engeligarail,	Algrail.
Tarreman,	Termet.
Falalis,	Falalis.
Futalis,	Faralies.
Lüsagä,	Otagu.
Falelegalä,	Falelmelo.

According to Kadu.

Falelemoriet.
Faleelepalap.
Faloetik.
Lollipellich.
Woesafo.
Lugalop.
Jesany.
Seliep.
Pügel.
Tabogap.
Tarrematt.
Piel, and
Ulimiré, the abode of Toua, the chief of the island-chain, and native place of Kadu.

Fatoilep (K.), Farroilep (C.), Farruelap (T.), and according to him in 8° 30′ north latitude, 144° 30′ east longitude. According to Cantova, it was seen, in 1696, by Juan Rodriguez, between 10° and 11° north latitude. A small and low group, consisting of three islands.

The bank of St. Rosa, near the south coast of Guahon, whose existence was particularly proved by Dampier, in the Cignet, in 1686, and again by Juan Rodriguez, in 1696, is not to be found any more; and in 1804 the Maria sailed over the place where it is marked on the chart.

Uetasich is, according to Kadu, a shoal to the north of Ulea, which may serve mariners coming from Feis as a mark not to miss Ulea. But Uetasich is not to be seen if you take the right course. The water is of a white colour. The sea has no surf.

Eurüpügk (K.), Eurrupuc (C.), Aurupig (T.), an inconsiderable low group of three islands, two of which are extremely small, at no great distance from Ulea. According to Kadu and Cantova they lie to the west, but according to Torres, to the south.

We think that the two islands, 1791, on Arrowsmith's chart, ought to be mentioned here, though they are very remote. Compare Sorol.

The four following form a chain, which according to C. lies to the east of Ulea, according to T. towards east south east, and according to K. towards sunrise.

Iviligk (K.), Ifelue (C.), Ifelug (T.), (the thirteen islands, or the two low islands of Wilson?) Low group of islands.

Elath (K.), Elato (C.), Elat (T.), (the two low islands of Wilson?) A small and low group, of which only the island after which it is named is considerable. There are four or five smaller ones.

Lamureck (K.), Lamurrec (C.), Mugnak (T.), Lamursee, in Krusenstern, likewise frequently called Lamurca, Lamuirec, or Falu, by Gobien, and also in Serrano's chart. (Swede's islands; the six islands of Wilson?)

Luyto (Krusenstern) makes the islands to be thirteen in number. A principal group of low islands: their names are Puc, Falait, (Falu, Serrano?) Toas, and Uleur, given on Cantova's chart, must relate to single islands of the group, and perhaps also Olutel, though they are laid down near Elato.

The Banc de Falipy, of Cantova, is mentioned neither by Kadu, nor by Don Luis de Torres.

Setoan (K.), Seteoel (C.), Satahual (T.) (Tucker's Island: Wilson, in 7° 22′ north latitude, 146° 48′ east longitude?) A large low island, lying by itself.

Ollimirau (K.), Olimarau (C.) A small low group, wanting on the chart of Don Luis de Torres. Kadu places the islands to the east of Setoan; Cantova, to the N.W. of Lamureck, half way to Fayo, a situation which must be incorrect, as it is never touched at on the voyage from Lamureck to Fayo and Guahon; and if our notion of Wilson's islands is correct, there is no room between Lamureck and the northern desert island for another group. We should look for Ollimirau to the east or north-east of Setoan.

Fayo (K.), Faheu (C.), Fallao (T.), and, according to him, in latitude 8° 5′ north, and 146° 45′ east longitude.* An uninhabited island, without any fruit-trees or sweet water, which only collects

* Fayo, according to that, would lie 43′ north, and 3′ west of Tucker's Island; and if the Swede's islands are Lamureck, the voyage from this group by Fayo to Guahon in two or three days must have been incorrectly divided, as you ought to reach Fayo in one day. We observed that the voyage from Fayo to Guahon, a distance of about seven degrees, or 360 miles, is performed in three days, or seventy-two hours, at the rate of five knots; that is, five miles an hour

in pits after the rain. Fatoilep, Ulea, Iviligk, Elath, Lamureck and Ollimirau, are visited for turtle and bird-catching.

Bigellé (K.), Piguelao (T.), and according to him in latitude 8° 6′ north, 147° 17′ east longitude, wanting in Cantova's chart. A similar island, likewise visited on account of the chase, from Elath, Lamureck, and Ollimirau.

Oraitilipú (T.), a shoal of twelve fathoms, between the two above-mentioned islands, in latitude 8° 6′ north.

Don Luis de Torres has determined another shoal of twenty-four fathoms, in 8° 20′ north latitude, 149° east longitude.

The islands hitherto mentioned form the second province of Cantova, which in his time was divided between the two states of Ulea and Lamureck, but now acknowledges the Tamon, or Prince of Ulea, as the sole master. This Tamon, of the name of Toua, is, besides, acknowledged in several of the more easterly islands, which Cantova reckons in his first province; and especially, according to Kadu, in Saugk, Buluath, and the high land of Tuch. According to Don Luis de Torres, when Toua dies, these islands will not fall to his successor in Ulea, and this maritime state will be divided.

The same language is spoken in all the islands of the second province of Cantova.

The intelligence respecting the eastern islands, which form Cantova's first province, Cittac, under the Prince of Torres or Hogoleu, are the most uncertain and dubious, and it is very difficult to throw light on their geography.

Kadu was never in these islands himself; he makes five groups, or islands, follow, always in a direction from Ulea, according to the rising sun, or rather inclining to the south.

Saugk (K.), Sog (T.), Scheug, or, according to the situation, Shoug (C.)? A low group.

Buluath (K.), Puluot (C.), Poloat (T.) A reef, of which only the island of this name is inhabited. Saugk and Buluath still speak the language of Ulea.

Tuch (K.), Rug (T.) Shoug, or, according to the situation, Scheug (C.)? This is the only high land which Kadu's information mentions to the east. Tuch has very high mountains, and, according to Don Luis de Torres, it has a peak. The inhabitants live in constant war with those in the distant islands, (Giep and Vageval). Their language deviates much from that of Ulea; Don Luis de Torres calls it a separate one. Kadu had had dealings with the inhabitants of Tuch and Buluath, at Ulea, where they brought tribute, and traded.

Savonnemusoch, and Nugor, are rich and low groups of islands, which Kadu places at a great distance in the same point of the compass; each is said to speak a different language. In the name Nugor we may recognize Magor (T.), Magur (C.)

Toroa, and Fanopé, are, according to Kadu, low groups of islands, which are very well known to the inhabitants of Buluath, as many ships from those shores are frequently wrecked there. Several of these strangers, after a short stay at Buluath, once tried to find their way back to their home. After wandering about for a month, they again landed at Buluath. The language of Ulea is spoken in Toroa and Fanopé.

One of the songs of these islanders, which Kadu learnt at Ulea, from the inhabitants of Buluath, preserves the knowledge of

Malilogotot, a very remote and low group of islands, with which they have probably been made acquainted by a boat wrecked from thence. A different language is spoken there, and the inhabitants are said to eat human flesh. (We are reminded of the Repith Urur of the Radackers.)

Wuguietsagerar is a dangerous reef, well known by the inhabitants of Buluath, after which they seem to direct themselves in their voyages. It is said to be at a considerable distance from this island. It forms a crescent, in which it is very dangerous to get entangled. The entrance must be avoided, and the whole reef left on one side.

Giep, Cuop (C.?), and

Vageval, are low groups of islands, at a great distance from Tuch, and at war with that island. Kadu has no further knowledge respecting them.

Lomuil, and

Pullop are the names of islands which he remembers to have heard of at Ulea. The chart of Don Luis de Torres coincides with that of Cantova, in the general arrangement of the islands of this eastern province, as well as in most of their names. When he first drew it he forgot the principal island Torres, or Hogoleu (C.), which is marked on Serrano's chart under the name of Torres, and which is not mentioned in Kadu's information. But after he had inserted in it the twenty-nine islands of Monteverde, (in S. Rafael, 1806), according to their stated latitude and longitude, where, in the circle that the province of Cittac forms, they filled pretty nearly the place which Hogoleu occupies in Cantova, the experienced seaman, Olopol, from Setoan, gave these islands the name of Lugulus, in which we must perhaps recognize Cantova's Hogoleu.

Cantova has nineteen islands; Don Luis only sixteen, with Lugulus. He wants those which in Cantova close the circle in the south-east, and are five in number, and in the other parts of the circle he has three new ones for one that is wanting: namely, —

According to Cantova.	*According to D. Luis de Torres.*
I. Torres, or Hogoleu, in the east, and from thence following the circle to the north.	I. Lugulus.

According to Cantova.	*According to D. Luis de Torres.*
2. Etel.	2. Pis (4. C.)
3. Ruac (4. T.)	3. Lemo.
4. Pis (2. T.)	4. Ruac (3. C.)
5. Lamoil (7. T.)	5. Marilo.
6. Falatu (6. T.)	6. Felalu (6. C.)
7. Ulatu (8. T.)?	7. Namuhil (5. C.)
8. Magur (9. T.)	8. Fallao (7. C.)?
9. Uloul (11. T.)	9. Mayor (8. C).
10. Pullep (12. T.)	10. Pisaras.
11. Puluot, or Legnischel, to the west, nearest to Setoan (14. T.)	11. Olol, to the west, next Setoan (9. C.)
12. Temetem (13. T.)	12. Pollap (10. C.)
13. Schoug (16. T.)	13. Temetem (12. C.)
14. Scheug (15. T.)	14. Poloat (11. C.)
15. Pata.	15. Sog (14. C.)
16. Peule.	16. Rug, in the south where the circle remains open.
17. Foup.	
18. Capeugeug.	
19. Cuop.	

The comparative view which the annexed charts afford, renders a further description unnecessary.

Cantova gives his province Cittac a different language, distinguished from that of Ulea. On the other hand, Kadu's testimony is of more weight, at least as far as respects Buluath and Tuch.

Cantova mentions, in a vague manner, a great number of islands far to the east of Cittac, of which he only names, and particularly distinguishes Falupet (Fanopé, K.?) The shark is said to be worshipped there. Navigators of these islands, who have been wrecked upon the more westerly ones, have made them known.

We now return to Ulea, and from thence, number the islands of the western chain.

Feis (K. and C.), Veir, according to the pronunciation of Radack, Fais (T.) and Pais, according to the chart of Serrano, seen by the Nassau fleet in 1625. It lies to the north-west of Ulea, and the voyage to it appears to be one of the most dangerous, requiring, according to Kadu's testimony, (to whom, however, in this instance, we do not give the most implicit credit,) fourteen days' time. Feis, though it is of the same formation as the other low islands, is higher, and by far more fruitful. Three islands, or territories, are called Litötö, Soso, and Vaneo. The chief of Litötö is independent prince of Feis.

Mogemug (K.), Mogmog (T.), Egoi, or Lumululutu (C.) (He gives the first name to the western islands of the group or leeward islands, and the others to the eastern or windward islands.) Los Garbanzos, on his improved chart, and called by F. Juan de la Conception, Ulithi in Eap, discovered by Bernard de Egui in 1712, the group to which Cantova went as missionary, and where he died. A main group of low islands, and seemingly larger than Ulea. It lies between Feis and Eap, at a little distance from both, and obeys its own chief.

Cantova has given the names of twenty-three islands: Kadu mentions twenty-six, among which the most of Cantova's may be recognized; namely,

According to Cantova.	*According to Kadu.*
Mogmog.	Mogemug.
Sagaleu.	Thagaleu.
Oiescur.	Essor.
Falalep.	Talalep.
Guielop.	Elap.
Gaur.	Cor.
Lusiep.	Lussiep.
Alabul.	
Pugelup.	Pugulug.
Pig.	Pig.
Faleimel.	Faleiman.
Faitahun.	Teitawal.
Laddo.	
Fantarai.	Fasarai.
Caire.	
Pigileilet.	Pigeleili.
Soin.	
Troilem.	
Lam.	Lam.

According to Cantova.	*According to Kadu.*
Elil.	Elell.
Petasaras.	
Medencang.	
Marurul.	Malauli.
	Tongrosz.
	Malimat.
	Tarembag.
	Song.
	Elipig.
	Eo.
	Eoo.
	Lasz.

Feis and Mogemug, according to Cantova, form the third province, and are said to have a language of their own. But the language of Ulea is spoken there with very little deviation.

Eap (K.), Yap (C.), Yapa (T.), Ala-cap, Account of the Pelew Islands, p. 21 in the note. It was seen by the Nassau fleet, in 1625; by Funnel and his companions, in 1705; and by the Exeter, in 1793, according to whose determination it is now laid down on the chart. A high and considerable island, which, however, has no very high mountains like the Pelew islands. It was formerly under one chief, and enjoyed peace; but now war rages between the petty chiefs of the several territories, of which Kadu named forty-six; namely,

Kattepar, Sigel, Sumop, Samuel, Sitol, Suomen, Palao, Runnu, Girrigai, Athebué, Tugor, Urang,

Maloai, Rumu, Gilifith, Inif, Ugal, Umalai, Sawuith, Magetagi, Elauth, Toauwai, Ngari, Gurum, Tabonefi, Summaki, Sabogel, Samusalai, Tainefar, Thorta, Unau, Taumuti, Sul, Sütemil, Täp, Ulienger, Wutel, Laipilau, Süllang, Thelta, Urieng, Meit, Feidel, Tumunaupilau, Sop, &c. Smaller islands along the coast of Eap are without either names or inhabitants.

Eap has a language of its own, which is now spoken only in the following group.

Ngoli (K.), Ngolog (T.), Ngoly (C.) A small low group, at a short distance from Eap to the south, and on the way to Pelli. It has only three islands, and of these only that after which the group is called is inhabited, and this has not above thirty people. The names Petangaras and Laddo, in Cantova, relate to the other islands of the group, and the name of Laddo is preferred in many of the new charts; for example, Burney's.

Between Eap and the Pelew islands we may compare, with Ngoli, the island de los Reyes Saavedra, 1528, de los Matalotes, Villalobos, 1542; those of Hunter 1791, and the islands seen 1796. Those of Hunter seem to correspond most with the situation of Ngoli. The islas de Sequeira, 1526, are, with much probability, supposed by Burney to be los Martires of the Spaniards, 1802, more to the west than the Pelew islands.

Pelli (K.), according to the pronunciation of Ulea, and, according to him, more correctly *Walau*;

Pannog (T.), Paleu and Palaos (C.), the Pelew islands of Wilson. Los Arrecifes of R. L. de Villalobos, 1542. Islands of Thieves of Sir Francis Drake, 1579?

An archipelago of high islands, divided into two rows, which are constantly at war. The Pelew islands are perfectly known to us, and are regularly visited by our ships. They have a language of their own, and even the people seem to differ in many respects from the Carolinians.

The chart of Don Luis de Torres is bounded here, and Cantova has nothing in addition, except St. Andrew's islands to the south-west of Palaos.

Kadu reckons besides, in this direction, Lamuniur (K.), Lamuliur (Father Clain). Compare the Doubtful islands of St. John.

Sonsorol (K.), and Relation et Lettres Edifiantes, tom. xi. p. 75., is also found on the accompanying chart. Sonrol in Cantova: both names in Fr. Juan de la Conception.

Kathogube (K.), Codocopuei (C.) The two latter are the St. Andrew's islands, where the first missionaries, Cortil and Duperon, were abandoned, in the year 1710, and which have since been forgotten. They appear in the reports of the mission as islands of one group; and Kadu, who separates them, and gives their distances from each other in days' voyages, cannot well be considered as authority for islands which he has never visited.

Wull (K.), Poulo and Pulo, after the reports of the missions, according to which it lies S. ¼ S.W. of Sonsorol. Compare Carteret's Current island.

Merir (K.), Merieres, after the reports of the missions, and, according to them, S. ¼ S. E. of Sonsorol. Compare Warren Hastings' island.

The names of the two latter islands, Pulo Maria and Pulo Ana, are noted on the chart to Fr. Juan de la Concepcion, t. ix. p. 150. Pulo Anna and Pulo Mariere, on other charts, are corrupt compounds of different languages. The Malayan word *Pulo*, for island, is well known to the Europeans in the Malayan archipelago.

All these islands to the S.W. of the Palaos, are low islands, or groups of islands, whose peaceable and inoffensive inhabitants speak the language of Ulea. The events at Sonsorol, where islanders from Ulea and Lamureck served as interpreters to the Spaniards, confirm Kadu's statement.

According to Kadu, the trading-boats go from Ulea to these islands, namely, as far as Merir, by way of the chain of the northern islands, as we have followed them from Ulea; but they return from Merir to Ulea by another way, viz. by

Sorol, or Sonrol (K.), (not the Sonrol of St. Andrew's islands), Zaraol (Cantova), according to whom it is under the dominion of Mogemug, and is fifteen leagues distant. It is marked on his chart, but the name is omitted.

A small low group of two islands to the south, and no great distance from Mogemug.

Compare Philip's islands of Captain Hunter, 1791, and the two islands, in 1791, which we have already mentioned, with more probability, at Eürüpugk.

Sorol, according to Kadu's statement, appears to have been peopled from Mogemug, and to have been under its jurisdiction. It is now wholly depopulated. These traditions also mention —

Lügülot, a low group of islands; a boat from which, bound to

Umaluguoth, a remote desert island, for the purpose of catching turtle, was driven upon Sorol. The strangers were guilty of robbery. The quarrel which ensued led to bloodshed. The chief of Sorol, and seven men and five women, on his side, were killed; on the side of the strangers, about four men. Subsequently, some of the inhabitants of Sorol went to sea, but never returned. At last only one man and some women remained on this group. We can offer a conjecture on the situation of these islands.

Don Luis de Torres has enabled us to look for the discoveries made by Wilson, on board the Duff, in 1797, among the Carolinas; and we are inclined to recognize Ulea in his populous and rich group of thirteen islands, though the number of the islands, among which he only counts six large ones, does not agree. If we are not mistaken in our supposition, the island-chain runs from Ulea to Setoan (the group of thirteen islands, and Tucker's Island), under the seventh degree of north latitude, from west to east, the direction which it takes on Cantova's chart, and not from W.N.W. to E.S.E. as Don Luis de Torres draws it. This chain occupies, besides, about three degrees of longitude, instead of extending above five.

The relative situation of the islands may be more easily obtained from the accounts of the natives, than their distances. The points of the compass may be pointed out with precision; the distances, according to the time required for the voyage, and even here, all measure of time is wanting. Cantova appears, in drawing his chart, to have commenced like Don Luis de Torres, from Ulea, which he properly marks to the south of Guahon. Both of them had fixed points for the western part, between which it only remained to place the other islands. This was not the case for the eastern part, where they had unlimited space open to them. We can only wonder at the accidental agreement of the standard which they have applied. If now we have a right to apply the reduced scale, which Wilson's discoveries furnish, to the province of Cittac, we must look for it between 148° and 152°. east longitude of Greenwich, and $5\frac{1}{2}$° and $8\frac{1}{2}$° north latitude. And we find, in fact, that several islands, discovered by our navigators, have been found in these limits; namely,

The island seen by Captain Mulgrave in the Sugar-Cane, in 1793, and by Don J. Ibargoitia, in 1801; which the latter (without assigning a reason) and Arrowsmith take for Quiroso, or St. Bartholomew, a large, tolerably high island, which Quiros discovered in 1595, after the death of Mendana. We observe that low groups of islands must be in the west, near Quiroso.

The island of Cota, in 1801.

A low island, seen in 1796.

Los Martires.

The Shoal of Don Luis de Torres, in the Maria, in 1804.

The Anonima of Espinosa's chart.

And the high land of M. Dublon, in the St. Antonio, in 1814.

The coincidence of Monteverde with Lugulus in the chart of Don Luis de Torres, is to be considered as an illusion. On the other hand, we are not indisposed to unite with Burney, Hogoleu, and Quiroso, but we believe that this island must be removed westwards from the place where he has laid it down, and where the low group of St. Augustin, of F. Tompson, in 1773, really lies. The situation of Dublon's Island, which is described, like Tuch, to have a high peak, seems to us to correspond with Quiroso or Hogoleu, whereas Ibargoitia recognizes Quiroso in an island, which appears to us to take up the place where we should have rather looked for Tuch.

To the east of Cittac, as far as to the island-chain of Ralick and Radack, remains a space of about fifteen degrees, in which the uncertain accounts of Cantova lead us to suppose many islands, and where our navigators have in fact discovered many. We merely observe among them, and even to the east, some high islands, such as Strong Island, (Teyoa of Arrowsmith) which is said to rise to a high mountain, and Hope, in 1807. St. Bartholomew Island of Loyasa, in 1526, lies more to the north, likewise a high land; to the west of which there are low islands. The islands seen by the Nassau fleet have been erroneously referred to these.

The boats from the province of Ulea and Eap, which are driven to Radack, convince us that the monsoon reaches much farther to the east than we have supposed.

The seamen of these islands, who find their way back from Radack again to their home, and on the other side go to and from the Philippines, show us that their navigation embraces a space of forty-five degrees of longitude, which is almost the greatest breadth of the Atlantic Ocean.

RADACK, RALICK, REPITH-URUR, BOGHA, THE CORNWALLIS' ISLANDS.

We had opportunities at Radack to examine more closely the formation of the low Coral islands, and to illustrate and correct our previous observations on this subject.

We conceive an island-group of this formation as a mass of rock, which rises with perpendicular walls from the unfathomable depth of the ocean, and forms on the surface an overflowed *plateau.* A broad dam, constructed by nature round the edge of this *plateau,* changes it into a basin. This dam, or reef, is generally on the side of the circumference turned to the wind, and at the time of ebb rises like a broad causeway above the water. On this side, and particularly on the projecting angles, the most islands collect on the back or ridge of this dam. On the lee side, on the contrary, it generally keeps under the water. It is there partially interrupted, and its intervals are sometimes broad enough to admit even large ships, which sail with the current into the internal basin. Within the inside of this entrance frequently lie single rocky banks, like fragments of a broken wall, or indications of it. Other similar banks lie scattered here and there in the interior of the basin. They seem to be of the same nature as the surrounding wall, but never project above the surface. The inner sea, or lagoon, in the more considerable group of Kawen, was from twenty-five to thirty-two fathoms in depth ; in the smaller one of Eilu, twenty-two fathoms, with numerous shoals. The bottom is finer or coarser coral sand, and in some places corals. The ocean is already coloured, at this depth, with the deep azure blue which distinguishes the pure waters of this ocean. The eye recognizes the shoals at a distance, and the lead may be dispensed with.

That part of the reef which projects above the water, and which may be examined, consists of almost horizontal masses of hard lime-stone, difficult to be broken, which is composed sometimes of finer, and sometimes of coarser fragments of madrepores, mixed with numerous shells and thorns of sea hedge-hogs, and breaks in large tables, which loudly sound under the hammer. The stone contains the lythophytes only as fragments, and no where in the situation in which they have grown and lived.

The surface of the dam is swept and smoothed, on its edge turned to the exterior sea, by the dashing of the surf. On the outer edge itself, against which the breakers rage, blocks of the stone have been torn up from their places. Such blocks are found again scattered here and there, on the side turned to the Laguna. This side is sloping, and the edge less sharply marked lies under the water. It seems that the beds are declivous in the interior, and that the upper beds do not reach so far as those on which they rest. The anchorages found in the Laguna of the groups, at four, six, and eight fathoms' depth, protected by the chief islands lying to the windward, are owing to this declivity of the stony layer. The lead, however, generally sinks at the inside, and along the reef, from a depth of two to three fathoms immediately to twenty or twenty-four, and you may pursue a line in which on one side of the boat you see the bottom, and on the other the azure-blue deep water.

A fine white sand of madrepore fragments covers the declivity of the dam, which is washed by the water. A few kinds of branching madrepores, or millepores, rise partially from this bottom, in which they have fixed themselves, with roots of a round form. Several others grow on the stone walls of larger clefts, the bottom of which is filled up with sand; among these also the *Tubipora musica*, which we saw in a living state, and the producers of which we recognized to be a polypus of the form of a star of eight rays. Species which cover the stone, or assume a lozenge form (*Astrea*) are always met with in the constantly-watered hollows of the bottom, next to the breakers. The red colour of the reef, under the breakers, is caused by a *Nullipora*, which covers the stone wherever the waves beat, and under favourable circumstances, assumes a stalactitical form. The colour and silky lustre, which disappear in the air, immediately decided us to ascribe to this substance an animal nature, and the treatment of the bleached skeleton, with diluted nitric acid, confirmed our opinion, which had been founded on analogy. The cursory view distinguishes, only by the colouring, and a certain velvet-like appearance, the lythophytes, with fine pores in a living state, from their dead bleached skeletons. We found only the *Millepora cærulea*, and the *Tubipora musica*, and a yellowish, red-brown *Distichopora*, with coloured skeletons; but never saw the latter alive. The kinds with larger stars, or *Lamellæ*, have larger and more distinguishable polypuses. Thus an animal, resembling the *Actinia*, covers the end-branches of a species of *Caryophyllia*, which we also found alive above low water-mark; the branches and roots seem to be bleached and dead. We often see in the lythophytes living branches, or parts existing with others that are dead; and the species, which otherwise assume a spherical form, spread out in places where sand is carried, into flat surfaces, with a raised edge, because the sand kills the upper part, and they can only live and grow on the circumference. The enormous masses of one growth, which are met with here and there on the islands, or on the reefs, as rolled pieces of rock, have been probably formed in the tranquil depths of the ocean. Above, under the influence of various agents, only masses of inferior size can be formed. A broad limbed *Corallina*, in a living state, has a vegetable green colour, which it loses when dried. There occurs only one small species of *Fucus*, which is not yet described, (*Fucus Radaccensis*, Mertens.*)

The sand which is deposited on the inner declivity of the reef there accumulates in places into banks. These sand-banks become islands. These, as we have already observed, are more numerous, of larger extent, and richer in soil on the windward side, and on the projecting angles of the group. Smaller, as it were infant, islands lie within upon the reef, and the inner sea constantly washes their beach. Some islands rest on stone layers, which slope towards the inner sea. These layers, where they should appear, towards the outer sea, are mostly covered with another layer of the same stone, which consists of coarser fragments of madrepores, and appears on its upper surface uneven and eaten away. This exterior layer is frequently broken to pieces, and lies in large blocks, displaced from its natural situation. In other islands we perceive, on the external and internal

* *Algæ*, which seem to be entirely wanting on the low islands, are found on the reefs at the foot of the high land. We collected on the reefs of Woahoo *Fucus natans* and others, several Ulveæ, &c.

side, a thin layer; the formation seems to be new, and layers of sand mostly alternate with those of limestone. This is always the case on the beach of the inner sea.

A dam of great madrepore-rolls cast up upon this foundation, forms, towards the breakers, the exterior edge of the islands. The interior of them consists of flats and small hills. Towards the beach of the inner sea, the ground is raised a little, and of fine sand. On the island of Otdia, in the group of the same name, the inner sea encroaches on the land in one place, and a *Lythrum Pemphis* grows with bare roots on the rocks which are washed by the sea. In the inner part of Otdia there is a lake of sweet water, and in Tabual, of the group of Aur, a marshy ground. There is no want of fresh water in the larger islands; it rises sufficiently in the pits dug for the purpose.

On the dam of fragments which surrounds the exterior of the islands, grow, first, *Scaevola Konigii* and *Tournefortia sericea*. These screening bushes gradually rise, and their thickly intertwined branches present outwardly to the wind an inclined surface, under the protection of which the wood or bushes of the interior of the island rises. The *Pandanus*, and where the soil is richer, also a *Cerbera* constitute the chief vegetation. *Guettarda speciosa*, *Morinda citrifolia*, *Terminalia Moluccensis*, are common on all the islands. *Hernandia sonora* is seldom wanting on the richer ones; *Calophyllum inophyllum, Dodonea viscosa, Cordia sebestena,* &c. are met with singly. On the northern, poorer groups, grow *Lythrum Pemphis* and *Suriana maritima,* on the beach of the inner sea, on the dry sand. They are wanting in Kawen and Aur. Only the bank of the inner sea is habitable for man, and there he builds his huts under the cocoa-trees which he has planted.

The Flora of these islands is poor: we found in the Radack chain only fifty-nine kinds of plants, including those, seven in number, which are found only in a state of cultivation. Three and twenty of this number, among which were five cultivated, we had already met with in Woahoo, and twelve, including the cocoa-tree, on Romanzoff Island, where we, upon the whole, collected only nineteen species. We afterwards met with about twenty of them in Guahon. We observed neither oranges nor cabbage-trees grow upon the Radack chain, as far as we were acquainted with it, productions which have been ascribed, upon dubious information, to the Mulgrave islands. *

We do not think that the Flora is confined to the number of plants above mentioned, but, on the contrary, that several kinds escaped our observation, even on the groups which we visited, and, of which, we were not able to explore all the islands;

* Vide the Voyage of Governor Phillips, 2d edit. London, 1790, p. 218.: the voyage of the Scarborough, Captain Marshall.

and that the southern groups in particular, which we have not seen (Arno, Meduro, and Millé), having richer soil and older vegetation, must produce more plants, which are entirely wanting on the northern and poorer islands. The vegetation of this chain seems to have begun in the south, and that man pursued its progress to the north.

Bygar, still desert and without fresh water, is only visited for the purpose of catching birds, and turtles. Udirick, a reef of inconsiderable circumference, and poor in land, has only two inhabited islands. In these the cocoa rises indeed above the rest of the wood, but the vegetation seems poor, and the bread-fruit tree rare. Tegi, near Udirick, desert and of a scanty verdure, is scarcely known by name among the people of Radack. Eilu (perhaps, more correctly, Eilug) is the poorest of the groups at which we landed. Udirick and Eilu obtain the aroma, a plant which they have not, from the more westerly group of Ligiep. The bread-fruit tree is wanting in Ligiep, and the cocoa-tree does not rise above the rest of the wood. Temo, half way to Ligiep, is a small desert island, where those who undertake the voyage pass the night. Mesid, a single detached island, lying eastward, about two miles in its greatest diameter, did not afford us, on the lee side, where we approached it, the prospect of any abundant vegetation. Only single cocoa-trees rise from its centre, and the fresh water which they offered us to drink was extremely bad. Yet it is distinguished from all the groups of Radack, which we have visited, by its more numerous population. We estimated the number of people, who were collected at our approach, on the shore and in boats, at a hundred at the least. The considerable group of Otdia, with which we became best acquainted, has scarcely so large a number of inhabitants, including women and children. Only one island of this latter group has lofty cocoa-trees, and, on this alone, there are many roots and traces of trees that have formerly perished. Erigup, near Otdia, is a poor, inconsiderable group, inhabited by only five men and some women. We found Kawen, the largest group seen by us, in older cultivation, and in a more flourishing state. The Flora is richer, and it was there we first discovered the plantain, which seemed to have been newly planted. The island of Tabual, the only one of the group of Aur at which we landed, had an uncommonly luxuriant vegetation. Behind a thick wood of lofty cocoa-trees, there are in the low grounds plantations of banana and arum, and some plants grow there, which are not to be found on the other groups. The southern groups of Meduro, Arno, and Millé are said to be richer in bananas and roots, and the two first are equal to all the rest of the chain in population and power. Limmosalülü, to the north of Arno, is a reef or rock over which the sea breaks, and which serves the mariners of Radack as a guide.

The sight of all these groups, and their single islands, has a tiresome uniformity. Looking from the outer sea, where the cocoa-tree is not seen rising above the rest of the wood, it would hardly be supposed that there were any inhabitants. From the inner sea you see the settlements and the progress of cultivation. Only one island of the group of Otdia is distinguished, and attracted our attention already, on the outer sea, by the appearance of higher land. It rose like a beautifully verdant hill above the mirror of the waves. This island occupies the projecting angle of the northern reef. Differing in shape from the other islands, it has less breadth and more depth, as it extends over a point which the reef forms to the inner sea. The currents of this sea produce on the beach, which it washes, a violent surf. What appears to be a hill is wood. A tree, which circumstances would not allow us to examine, attains there an astonishing height and thickness, on a low ground of large masses of Madrepores. On other islands, where it is also found, it does not reach a considerable height. Trees which have been thrown down have frequently changed their roots into stems, their branches taking root, a circumstance which is not uncommon in Radack, and is an indication of hurricanes. The low wood towards the edge of the island appears to show its progressive enlargement. The pandanus is a stranger: nothing attracts man to this place. A sea-swallow, *Sterna stolida*, builds in countless flocks in the high airy summits.*

The most useful plant of this island-chain is the common pandanus of the South Sea islands (*Wob*). It grows wild on the sterile sand, where vegetation commences, and fertilizes the ground by the many leaves which fall from it. It luxuriates in the moist, low ground of the more fruitful islands. It is also diligently cultivated; numerous varieties, with improved fruits, which are to be ascribed to cultivation, are propagated by layers. Their seed produces the original species (the *Eruan*).† The fruit of the pandanus constitutes, in Radack, the

* At Erigup we saw the same birds in equally large flocks, hovering over one of the islands, which is otherwise not distinguished from the rest.

† There are above twenty varieties, which are distinguished by the exterior form of the fruit, or compound stone-fruit which they compose, and by the number of simple fruits or kernels which they contain. The masculine tree is called *Digar;* the wild growing feminine *Eruan:* varieties are, *Buger, Bugien, Eilugk, Undaim, Erugk, Lerro, Adiburik, Eideboton, Eromamugk, Tabenebogk, Rabilebil, Tumulisien, Lugulugubilan, Ulidien, &c.* (The fruit, which we received in 1816, from Udirick, was *Lerro;* the pandanus on the Romanzoff Island, *Eruan.*) That part of the fruit, from which the people of Radack and Ralick draw their principal subsistence, is used in the Sandwich, Marquesas, and Friendly islands, for aromatic shining yellow wreaths. We observe, by the way, that the species *Pandanus* requires a farther and difficult investigation, because the characters which most botanists have chosen to distinguish the species, which they have enumerated, are insufficient. Loureiro, Flor. Cochin, expressly observes, that the fruit of the *P. odoratissimus* is unfit to be eaten.

food of the people. The compound fibrous stone-fruits which compose the conical fruit, contain a spicy juice at their basis, the point where they are fixed. To obtain this juice, the fruit is first beaten with a stone, the fibres chewed, and pressed in the mouth. The fruit is also baked in pits, after the manner of the South Sea, not so much to eat it in this state, as to prepare mogan from it, a spicy dry confectionary, which is carefully preserved as a valuable stock for long voyages. To prepare the mogan, all the members of one or more families are employed. From the stone-fruits, as they come out of the baking-pit, the condensed juice is expressed by passing them over the edge of a shell, then spread out on a grate, covered with leaves, exposed over a slight charcoal fire to the sun, and dried. The thin slices, as soon as they are sufficiently dried, are rolled up tight, and these rolls then neatly wrapped in the leaves of the tree, and tied up. The kernel of this fruit is well tasted, but difficult to be obtained, and is often neglected. From the leaves of the pandanus the women prepare all sorts of mats, as well the square ones with elegant borders, which serve as aprons, as those which are used as ship's sails, and the thicker ones for sleeping upon.

After the pandanus, the cocoa-tree (*Ni*) holds the second rank. It is not only rendered valuable by its nut, which affords drink and food, vessels and oil for domestic use, but also and principally for its bast, from which cords and strings are made. On the pandanus depends the food, and on the cocoa-tree the navigation of these people. The manufacturing of the ropes and strings is an employment of the men, and even the greatest chiefs are seen engaged in it. The fibres of the bast are separated and cleaned, by maceration, in pits of fresh water. The rope is spun at the same time with the two threads of which it consists, equal bundles of fibres prepared before-hand being added to each. The wood of the old cocoa-tree, rubbed to powder, and mixed to a dough, with the help of the juice of the husk of the unripe nut, is boiled in cocoa-nut shells, or roasted over the fire, and used for food. Cocoa-nut shells are the only vessels in which the people can carry water about with them; they are preserved in longish woven baskets, made expressly for the purpose, with the eye upwards, and strung together. The cocoa-tree is planted and increased every where on the inhabited and uninhabited islands; but, notwithstanding the many young plantations with which you meet, it is only seen bearing fruit on the inhabited islands, and only on a few, and in the southern groups waving its crown high in the air. The cocoa-tree bears only very small nuts in Radack.

The bread fruit-tree (*Mä*) is not common in Radack, it is found only planted in the damp ground in the interior of the inhabited islands.

Old trees are however found even in some of the poorer ones. Its wood, as well as its fruit, is valuable, being employed for the keel of their boats; the other planks are made of drift wood. They are joined together with strings of cocoa-bast, and the joints caulked with pandanus leaves. The bread-fruit tree also produces a resin, which is used for various purposes. The bread-fruit tree, like all cultivated plants, has several varieties. The only one that occurs here deviates very little from the orginal form; its fruit is small, and the kernels often perfect.

A useful bast is procured from the bark of three different species of plants found here in a wild state. The principal is a shrub of the family of nettle (a *Boemeria?*) the *Aromä*, which grows only upon richer and moister soil.

The *Aromä* produces a white thread of uncommon fineness and strength. The Atahat (*Triumphetta procumbens*, Forst.) is a creeping plant; it is common, and, with the *Cassyta*, covers the most sterile sand. Men's aprons are chiefly made of its brown bast; they consist of loose strips of bast, which are sewed to a girdle of mat. It is also used for the ornamental borders in the finer mats. The fine white bast of the *Hibiscus populneus* (*Lo*,) which we found in Radack on the Aur group, is used for the same purpose. In the Sandwich islands, and other places, strings are made of this bast.

From the bulbous roots of the *Tacca pinnatifida*, which is very frequent here, a nourishing flour is procured, but which appears to be seldom prepared, and little used.

Three species of *Arum* (*Caladium*, A) *Esculentum*, *Marcrorhizon*, and *Sagittifolium*, the banana, and the *Rhizophora gymnorhiza*, are singly cultivated here and there, on different groups and islands. We found the banana first cultivated in Kawen, and saw it bearing fruits only in Aur. These species of arum do not find here the deep marshy ground which they require to form their roots, and are not calculated, in these islands, to constitute the essential food of the people.

Besides these plants, two of the rarer kinds, which are found wild, are generally planted round the habitations; two ornamental plants, a *Sida* and a *Crinum*, the sweet-scented flowers of which are mixed with those of the *Guettarda speciosa*, the *Volcameria inermis*, and, in Aur, *Ixora coccinea?* in elegant wreaths, which are worn in the long hair, which is fastened up, and in the ears. The people of Radack, though poor, are distinguished for their love of perfumes and elegant ornament.

The sea throws up on the reefs of Radack the trunks of northern firs and trees of the torrid zone (palms, bamboo). It provides the inhabitants, not only with timber for boats, but it also brings them, in the wrecks of European ships, the iron which they want. We did not find among them any tools to work their wood, except the valuable metal obtained in this manner; and when we would not credit the assertion of our friends in this respect, we found ourselves such a piece of wood, with nails remaining in it on the shore, under the lee of one of the islands of the group of Otdia. They receive, in a similar manner, another treasure, hard stones fit for whetting. They are sought for in the roots and hollows of the trees which the sea throws up: iron and stones belong to the chiefs, to whom they must be delivered, on payment of a reward; punishment being inflicted, in case of concealment.

The sea brings to these islands the seeds and fruits of many trees, most of which have not yet grown there. The greater part of these seeds appear to have not yet lost the capability of growing, and we have frequently entrusted to the bosom of the earth the present intended for it. We have collected them, and found among them, fruits of pandanus species, which are only met with in the larger countries lying to the west, those of *Baringtonia speciosa*, the *Aleurites triloba*, and other trees belonging to the general Flora of Polynesia, and which we afterwards saw in the west, on the Mariana islands. The greatest part of these seeds belong to the arborescent or creeping siliquose plants, which are found every where in abundance between the tropics. The seeds of the *Guilandina bonduc* are frequently found among them; and we met with the plant itself only once in the group of Otdia, on an island to the leeward. We observed that the seeds which, being cast by the tide over the reef, reach the inner, or lee side of an island, find more protection, a better soil, and circumstances more favourable for their growth than those which the surf throws upon the outside.

We frequently find rolled pumice-stones among the objects cast up by the sea, and compact masses of the *Cassyta*, resembling those which the *Zostera marina* forms on some of our coasts, and which are called on the French coast, in the Mediterranean, *Plotte de mer*.

Besides the *Mammalia*, which the sea nourishes, the dolphins, which the people of Radack kill only singly, and very seldom, as they are not numerous and powerful enough to surround them like other islanders in herds, drive them into their reefs, and catch them, the *Cachelot* *, and the more rare whales, only the rat, which is everywhere spread, is found in Radack, and which, as it has no enemy, has increased in a dreadful degree. Kadu, who seems to think that the rat is only to be found in the company of man, affirms that there are none in Bygar. In the more populous groups, for instance, in Aur, these troublesome animals are

* We saw, in the year 1817, a *Physeter macrocephalus*, near Radack.

sometimes pursued. They are collected together by means of baits, which are half surrounded by fire-pits, and driven into the fire, which has been lighted for them. The rat is eaten by the women in Udirick, and our sailors also saw women eating them in Otdia.

Fowls are found in Radack in a wild state; they are not used for food, except in Udirick: in the the other groups they are caught singly, and tamed for pleasure, without deriving any advantage from them. Here and there you find a fowl round the habitations, which, fastened by his foot to a string, and tied to a pole, reminds us of the Tagalese fighting-cock. A small white heron is likewise tamed. Besides the fowl and the pigeon of the South Sea, (*Columba australis,*) there are only wood and water birds, and these are not in great numbers, in the inhabited groups. The *Sterna stolida* is the most frequent, and fond of being in the neighbourhood of the surf.

The sea-turtle is caught at Bygar; of the class of *Amphibiæ*, there are, besides, four small kinds of lizards at Radack.

The Lagunas in the interior of the island-groups, are poor in fish. Outside, about the reefs, and at the entrances, there are swarms of sharks; but they seldom penetrate into the inner sea: it is said that these animals do not attack men at Bygar. We caught bonétos at the entrance of Eilu. The flying-fish is the most frequent in the neighbourhood of low islands. The Radackers catch it by torch-light, in the night. There are several sorts of fish which are not eaten, and are considered as poisonous. Kadu mentioned instances of persons poisoned in this manner. These species are eaten at Ulea, after a certain internal part (the liver?) has been taken out; and several (for instance, the *Diodon* and *Tetrodon* species) are even considered as dainties. Among the poisonous fish at Radack, two rays (*Raja*) are mentioned, which attain an extraordinary size; one of them has, like the *Raja aquila*, and *R. postinaca*, a large thorn on the tail; the other has five. Both, according to Kadu, dart these thorns from them in their defence; and when they are lost, they grow again within twenty days. They are always taken hold of before. They are pursued on account of their skin, which is used to cover drums. Both species are eaten at Ulea.

There is a great variety, both of univalve and bivalve shells. Many of them are eaten, and the shells of different ones are variously used. The Triton's horn is used as a signal trumpet. The *Chamagigas*, and other large bivalve shells serve as vessels, and are also used as instruments for cutting: the mother-of-pearl is sharpened into knives; and smaller kinds of snails are worn as ornaments, in elegant rows, round the head and neck.

Among the crabs, we noticed several small

species of *Pagurus*, which, in the borrowed speckled shells of all kinds of sea snails, go into the interior of the islands to seek their food.

The varieties of naked *Moluscæ*, worms, and zoophytes, are particularly numerous. We observed a cuttle-fish, several beautiful species of sea hedge-hogs and sea-stars, some *Medusæ* (but these not in all the groups,) and a few *Holothuria*. The poor inhabitants of Radack, who are often in distress for food, have on their reefs, in abundance, one of the animals, (*Trepang*,) after which the Chinese epicures are so eager, and frequently suffer want, without attempting to satisfy their hunger with these disgusting worms. The sea frequently throws on the reefs a small physalis, (*Physalis pelagica*, Tiles.) A worm which pierces the rocks below the high water mark, and lives in the inside of the lime-stone, and our common worm, are natives of these remote islands.

There are here but very few insects; we observed the *Scolopendra morsitans*, and the *Scorpio australasiæ*, of which the natives did not appear to be afraid; and the sting of which, according to Kadu, produces a local swelling, which is of short duration.

The inhabitants of Radack are neither of large stature, nor remarkable bodily strength. Though slender, they are well built and healthy, and appear to attain a very old age, accompanied with a considerable share of cheerfulness, and activity.* Children are suckled for a long time, and receive the breast when they can already walk and talk. The Radackers are of a darker colour than the people of Owhyee, from whom they are advantageously distinguished by greater clearness of the skin, which is not disfigured, either by the effects of the kava, or by any cutaneous disease. Both sexes wear their long beautiful black hair neatly and elegantly tied up behind. The children have it hanging down, unconfined, and curly. The men suffer their beard to grow, which is long, though not particularly thick.† Their teeth are generally spoiled by the nature of their food; from the chewing of the woody, fibrous fruit of the pandanus; and, sometimes, in the front, are broken off. It is less frequent with the chiefs, for whom the juice of the fruit is generally pressed out, and separated over the edge of a shell. Men and women wear, in their pierced ear-lappets, a rolled

* We must mention a natural deformity, which we observed in several wives of the chiefs of different groups, and in a young chief of the group of Eilu: it respects the fore-arm. The *ulna* appears, in the bend of the hand, dislocated above, and the fore-arm, which is bent, or more or less checked in its growth, is in some cases scarcely a span long; the hand is small, and bent outwards. A child in Otdia, had a double row of teeth in its mouth. We met also with an instance of a person deaf and dumb.

† We were told of a contest in Tabual, in which a man from Meduro was killed, whose beard reached to his knee!

pandanus leaf. The roll for the men is three inches in diameter, and for women, only half. It is, sometimes, covered by a very thin plate of tortoise-shell. Some older people had, besides, pierced the upper edge of the ear, to put flowers through it.

The skilful, elegant tattooing * differs according to the sex; in each it is uniform. For the men it forms over the shoulder and breast a triangle pointed to the navel, which consists of several variously combined stripes: similar well-disposed horizontal stripes occupy the back and the stomach. With the women only the arms and the shoulders are tattooed. Besides this regular designing, which is only executed when they grow up, and is wanting in very few, they have all, when children, groups of designs or stripes tattooed over their hips and arms, but more seldom in the face. Among these drawings we sometimes observed the figure of the Roman cross. † The place tattooed is very dark, drawn sharply and raised above the skin.

The dress of the men consists in a girdle with bast strips hanging down, to which is often added

* In the spring of 1816 we overlooked this tattooing in Udiruk (Kutosoff's islands.)

† Natives of the Mulgrave islands who went on board the Charlotte had a cross suspended from their neck, after the custom of the Spaniards. We did not meet with this ornament in Radack, and endeavoured in vain to discover in the mark mentioned any relation to Christians and Europeans.

a smaller square mat as an apron; boys go quite naked till they have arrived at manhood. The women wear two longer mats, fastened with a string over the hips, the girls wear, very early, a small apron. The men frequently have, besides the flower and shell wreaths, with which both sexes adorn themselves, a necklace of stringed dolphin's teeth, with plates of the bones of the same animal, or of the turtle, hanging in front. For this ornament thin, round shells, or slices of cocoa-nut shells, are used. We likewise found among their ornaments the tail feathers of the tropical birds, the feathers of the frigate bird, and bracelets formed of the shell of a large univalve, and well polished.

The Irus, or chiefs, are frequently distinguished by their large stature, never by immoderate corpulency.* The tattooing in them generally extends over those parts of the body which are not tattooed in common people, the sides, the hips, the neck, or the arms.

The houses of the Radackers consist only of a roof supported on four posts, with a hanging floor. They are only high enough to admit a person to sit under them. You climb through a square opening into the upper room which contains all their little property. They sleep on this floor or in the open space below, and several open huts of

* The chief of the group of Ligiep is said to make an exception, and to be a particularly fat man.

the form of a tent round about, serve as separate sleeping apartments. The roofs are of cocoa or pandanus leaves; the floor is strewed with very fine fragments of coral and shells, which are found on the shore. Only a coarse mat serves for a bed and a block of wood for a pillow.

At first we did not take these houses, which we often found deserted, to be the constant residences of the people. These seamen sail in their ingenious boats * with their family and all their goods from one island to another, and in this manner, after we had become intimate with them, the greater part of the population of a group always assembled near us.

The wild pandanus appears to be general property. A bundle of leaves of this tree (tokens of property) tied to the branch on which the fruit is ripening, secures the right of him who has discovered it. We have frequently, and particularly in the northern poorer groups, seen this fruit, which is the only food of the Radackers, devoured quite unripe. Cocoa trees are private property. Those in the neighbourhood of the habitations, the fruit of which is ripening, are often seen with a cocoa leaf fastened round the stem, which is intended, by the rustling, to give notice if any one should attempt to climb the

* The author of these essays leaves it to more competent persons to describe scientifically these vessels, which agree in their essential particulars with the often-mentioned proas of the Mariana islands.

tree. In the populous groups of Kawen and Aur orchards are frequently surrounded with a cord instead of a fence.

Besides providing for their subsistence, our friends have no other employment but their navigation and singing. Their favourite, their only property, are their boats and their drum, which are their play-things in childhood. It is particularly in the evening, when they assemble in a circle round their bright fire, that they sit singing their favourite songs. Intoxicating joy then seizes every one and all join their voices in the chorus. These songs resemble those of the Owhyeeans, but they are ruder and more discordant; the gradually rising tones of the song degenerate at last into a scream.

It was first and principally in the group of Otdia that we became acquainted with the amiable people of Radack. These people, who met us with friendly invitations, seemed for a time to fear us, in the conviction of our superiority. The chiefs always showed the most courage and the greatest confidence. Familiarity never made our friends importunate or troublesome. The comparison of our immense riches and their poverty, never degraded them to begging, seldom seduced them to theft, and never made them violate trust when it was reposed in them. We daily wandered through their islands unarmed, slept with our treasures, knives and iron, by our side, under their roofs, went upon long excursions in their boats, and confided in their character, as people confide among us, in the protecting vigilance of the law. At their own voluntary request we exchanged our names Whenever we appeared, the people came hospitably to meet us, and presented us with cocoa-nuts. We did not barter at Otdia, we made presents and received them in return. Some appeared to take as much pleasure in giving as ourselves, and, with great delicacy, brought us presents when none could be expected in return. Others behaved with more regard to self-interest. Where unheard-of events bring on relations never thought of before and custom is silent, the real character of men must show itself without disguise. The women conducted themselves with modesty and reserve, they retired when they first saw us, and appeared only under the protection of the men. In return for our small presents of rings, and glass beads, which they seemed to value less than odorous splinters of the wood of English black lead pencils, they presented us, in a graceful manner, with the ornaments which they then wore, their shell and flower wreaths. No woman of Radack ever came on board our ship.

Everywhere we met the picture of peace among an infant people; we saw new plantations, advancing cultivation, many children growing up, with a small population; the affectionate attention of the fathers for their offspring, pleasing unaffected manners, equality in the intercourse between chiefs and other men, no servility to the more powerful, and with greater poverty and less self-assurance, none of those vices which disgrace the people of the more Eastern Polynesia.

We first learned at Aur, that these indigent people also have their wars, and that the lust of dominion and conquest has extended its curse to them. They called upon us with our formidable iron (we had not made them acquainted with the more destructive powers of our arms) to interfere like the powers of fate in their bloody feuds.

The powerful Lamary set out from Meduro, to subject, by arms, all the northern island-groups of Radack. He governed now over Aur, Kawen, and the north of the chain, and had his residence in Aur. The people of Meduro and Arno wage war against him and his kingdom. Their expeditions in thirty boats, each manned by six to ten men, have extended themselves to Otdia.

The late battle in Tabual cost the lives of four men, three on the side of Meduro, and one on the side of Aur. In a former expedition, about twenty were lost on each side, on the same island.

Lamary, in the beginning of the year 1817, visited the islands in his territory, to assemble his fleet of war-canoes, likewise consisting of thirty boats, at Aur, whence he intended to proceed against Meduro. We expected to find this prince in Eilu; he was already at Udirick, at which group he visited us in his boat in the open sea. When we returned, towards the end of the same year, to Otdia, the expedition was assembled in Aur. Lamary had missed the island of Mesid, and being carried out of his course to other groups, had given up the reinforcements which he had expected there.

We will relate, at length, what we have learned of the religion, and social order, the manners and customs of our friends.

The inhabitants of Radack adore an invisible God, in heaven, and offer him a simple tribute of fruits, without temples and without priests. In their language, *Iageach* signifies God: the name of the god is *Anis*. When war or any other important affair is to be undertaken, solemn offerings are made, always in the open air. One of the assembly, not the chief, consecrates the fruits to the god, by holding them up, and invocation; the form is, *Gidien Anis mne jeo*, the assembled people repeat the last words. When the father of a family goes out on the fishery, or undertakes any thing important, he offers among his family. There are, on several islands, holy trees, cocoa-palms, into the crown of which *Anis* descends. Round the foot of such a tree four beams are laid in a square. There does not appear to be any prohibition to enter the space enclosed, and the fruits of the tree are eaten by the people.

The operation of tattooing is associated in

Radack with religious ideas, and cannot be undertaken without certain divine tokens.* The persons who desire to be tattooed pass the night in a house, on which the chief, who is to perform the operation, invokes the god; an audible tone, or whistle, is said to give his consent. If the token does not appear, the operation is not performed. Hence some persons never undergo it. If they were to transgress in this respect, the sea would inundate the island, and all the land be destroyed. The sea alone threatens these islands, and religious faith often suspends this rod over man. Against this, however, conjurations prevail. Kadu saw the sea rising at Radack, as far as the feet of the cocoa-trees, but it was conjured in time, and retired within its own limits. He named to us two men and one woman, at Radack, who understood this conjuration.

The desert island-group of Bygar has its peculiar god. The god of Bygar is blind, he has two young sons, of the name of Rigabuil, and the people who visit Bygar call each other, as long as they stay there, Rigabuil, that the blind god may take them for his sons, and do good to them. *Anis* must not be called upon in Bygar, the god would punish

* Our friends always refused, on various pretences, to confer this ornament on us. They often represented to us the serious consequences, the swelling of the limbs, the severe illness. A chief of Aur once appointed one of us to pass the night at his house, that he might tattoo him the following morning; but when the morning came, he repeatedly evaded the importunity of his guest.

the culprit with severe illness and death. Under a tree, in Bygar, offerings of fruits, cocoa, &c. are brought. Certain forms of conjuration are efficacious and infallible to make the water come into the pits; if the result is unfavourable, something has been forgotten, and the words were not said right. It is every where as with us.* At Bygar the sharks are not allowed to do any harm to man: the god will not suffer it. The inhabitants of all the groups of Radack visit Bygar by way of Udirick, only those of Eilu are not permitted to do it immediately. They must remain a month at Udirick before they sail there, and on their return must spend another month in the same group before they partake of the provision brought with them. This provision consists in the flesh of birds and turtles, which they first bake, and then dry in the sun. The use of salt is unknown at Radack.

The marriages, the funerals, the feasts upon several occasions seem to have no reference to religion. We were not happy enough to come to an understanding with Kadu respecting an existence after death.

Though no particular marks of respect are shown to the chiefs, they, however, exercise an arbitrary right over all property. We, ourselves, even saw chiefs, to whom we had made presents, conceal

* For example, the belief in medicine, the last thing to which incredulity still clings.

our gifts from the more powerful. They appear to be subordinate to each other in several degrees, without our being able rightly to understand their respective relations. Rarick was the most considerable in Otdia; his father Saur-aur, perhaps the real chief of the group, lived at Aur. Rarick and his son, a boy about ten years old, alone wore several stripes of pandanus leaves, in which knots were made, round the neck; and it seems to be a privilege. We have seen similar strips hanging in the houses of the chiefs, which, with dried fish-heads, unripe cocoa, and stones, have the appearance of consecrated objects. The succession is not direct from father to son, but from the elder brother to the younger, till after the decease of all, the first born son of the elder brother succeeds. Women are excluded, When a chief approaches an island a signal is given from his boat, and his wants are immediately supplied with the best that is to be procured. This signal is given by the person who is in the fore part of the ship, by raising his right arm, and calling. This was also observed when the officers of the expedition sailed in the boats of the natives. The chiefs are distinguished by freer motion in their gait, which the common man is not allowed to imitate.

The princes call their people together for war. The chief of each group joins the squadron with his boats. They undertake to surprise an enemy's group with combined force, and land. They never fight but on shore. The women take a share in the war, not only to defend their own country against the enemy, but also to attack; and in the squadron they form, though in smaller numbers, a part of the military force. The men stand foremost in the battle. Their weapons for distant combat are the sling, in the use of which they are not skilful, and a staff, pointed at both ends, which being thrown in the arc of a circle, turns round upon its centre, like the diameter of a wheel in motion, and penetrates with the end on which it falls. For close combat the dart, a stick of five feet in length, which is pointed, and furnished with shark's teeth or barbs. We saw only at Mesid a short crooked wooden sword, both sides of which were provided with shark's teeth. The women, unarmed, form a second line: some of them beat the drum, according to the word of command, first in slow and regular time (*Ringesipinem*), when the combatants, at a distance, exchange dart for dart, then in double quick time (*Pinneneme*), when they are engaged in close combat. The women throw stones merely with the hand: they assist their lovers in the contest, and throw themselves as deliverers and peace-makers between them and the vanquished enemy. The female prisoners are well treated: men are not made prisoners. The man adopts the name of the enemy whom he has killed in battle. Conquered islands are robbed of all their fruit, but the trees are spared.

Marriages depend on a free convention, and may be dissolved as they are concluded. A man may have several wives. The woman is the companion of the man, and seems freely and voluntarily to submit to him, in a just relation to the head of the family. In their wanderings the men go on before, as protectors, and the women follow. When any subject is discussed, the men speak first: the women, when called upon, take a share in the conversation, and attention is paid to them. In the time of peace they have no other employment than what we call women's work. The drum, which awakens delight in all, is in their hands. Unmarried women enjoy their freedom, under the protection of decorum. The girl requires presents from the man; but the veil of modesty is drawn over all the connections which unite the two sexes. We observed that the customary caress by touching the noses, even among men in the Carolina islands, as well as in those of the Eastern Polynesia, was usual at Radack only between man and wife, and in the shade where familiarity is concealed.

The bond of exclusive friendship between two men, which is found in all the islands of the first province in Radack, obliges the friend to give his wife to his friend, but does not bind him to seek for revenge by blood.

We mention with hesitation and abhorrence a law, the reason for which Kadu ascribed to urgent want, and the sterility of the niggardly earth. Every mother is allowed to bring up only three children; her fourth and every succeeding one she is obliged to bury alive herself. The families of the chiefs are not subjected to this barbarity. Natural children are brought up in the same manner as the legitimate. When they are able to walk, the father takes them to himself. When no father recognizes the child, the mother keeps it: when the mother dies, another woman takes care of the child.

The corpses of the deceased are entirely wound round with strings, in a sitting posture. The chiefs are buried on the islands. A large square space, surrounded with large stones, marks the place, under palms, on the inner shore. Those of the people are thrown into the sea. They treat in the same manner their enemies fallen in battle, according to their rank. A staff fixed in the ground, with annular incisions, marks the grave of the children who were not allowed to live. We have ourselves seen both kinds of graves.

A long time ago, an European ship was seen at Kawen, and remained a day in the neighbourhood of this group, without attempting to land. The chief, Saur-aur, our hospitable friend at Tabual, went on board this ship. (We observe that his name was at that time Laelidjü, he having since received his present name by an amicable change with a chief of the island chain of Ralick, who is now called Laelidjü, after him.) The natives procured, in exchange, iron and pieces of broken glass from this ship.* Kadu possessed himself of two of these pieces when he was at Aur, and recollected them on seeing some similar pieces among us, which he collected for his friends. No song has preserved the name of this ship; no names are snatched from oblivion.

We are the first Europeans who have landed at Radack and become acquainted with these amiable people. From principle and from inclination, from real sincere love, we endeavoured to neglect nothing that we could do for this people. On our first visit we put our friends in Otdia in possession of hogs, goats, and tame fowls: yams were planted, and melons and water-melons had sprung up, and throve well. On our return, after a few months, we found the place of the garden in the island of Otdia waste and desolate; not even a foreign weed remained to testify our good intention. The hogs had died of thirst, the fowls were not to be found; Prince Lamary had carried the goats to Aur, and had also transplanted there, from the May he, the benefactor of an amiable people,

* Wood may be scraped with pieces of glass; and they use them almost as we do the plane. They have a real value.

island of Otdia, the yams, which alone had escaped the hostile rats. The old chief, Laergasz, had discovered some yams planted by us on an island under his dominion. He found this root well-tasted, and, after he had eaten it, carefully replanted the leaves. This process, which is observed in the cultivation of the taro, had deceived his expectations.

The proper object of our second visit was to be beneficial to our friends. We brought them goats, hogs, dogs, cats, tame fowls, potatoes from the Sandwich islands (*Ipomoea tuberosa*, Lour. Coch.), the melon, the water melon, gourds of different kinds, those of which the fruit is used for valuable vessels, and others of which the fruit is eaten; the sugar-cane, the grape, the pine-apple, the apple-tree of the Sandwich islands (not an *Eugenia*), the tea-root (*Dracaena terminalis*), the lemon-tree, and the seeds of several useful plants on the Sandwich islands; the *Aleurites trilobà*, the nuts of which are used as tapers, and produce oil and stuff for dying; two of the shrubs, the bast of which serves to manufacture stuffs, &c.

We have carefully sowed the seeds which our friend Kadu has undertaken to attend to.

May Kadu proceed in his fine vocation with wisdom and energy! May he succeed in effecting what, without him, could not be hoped! May his good heart produce the good which he wishes!

establish its prosperity, conduct it peaceably and liberally to a better situation, and soon induce it to renounce a law revolting to nature, which was founded only in necessity!

We must confess that our friend stands alone exposed to the envy of his equals, the avidity and power of his princes, and the treasures which our love has heaped upon him, may gather the tempest over his head. But our apprehension can go still farther: the real treasure in iron, which we lavished with pleasure upon Radack, may kindle a destructive war between the south and the north of this chain, between it and Ralick, and blood be the fruit of our liberality.

The poor and dangerous reefs of Radack have nothing to attract the Europeans, and we congratulate our simple friends on remaining in their obscurity. The amiableness of their manners, the charming modesty which adorns them, are blossoms of nature, which are not derived from any abstract idea of virtue. They would show themselves equally docile to our vices; and, as the victims of our passions, would draw down upon them our contempt.

Ralick, near to Radack, in the west, is a similar chain of low groups of islands, with the geography of which even the women in Radack are familiar. Ralick is more fertile and populous than Radack. The people, the language, the tattooing are the same. No children are murdered; the women do not accompany them into war. The people are more prosperous and better fed than those of Radack: they wear a still larger ornament in their ears. Several men are mentioned by name, who were able to draw the extended ear-holes over their head.

Between the two island-chains, hostile and amicable relations take place. A chief of Eilu showed us the scars of wounds he had received at Ralick. Ralick employed fifty boats in a war against Radack; the chiefs of Radack went over, and a friendly intercourse was again established.

A European ship once came to Ralick. The ship is said to have remained at anchor in Odja (a chief group in this chain) for a long time (a year it was said.)

We suppose that likewise at Ralick the southern groups are the richest. Bananas roots, &c. are not found on all the groups.

Repith-Urur is represented to us as a considerable group of low islands, known to the inhabitants of Radack by the many boats cast on their reefs from that group. The boats and costume of the people of Repith-Urur are the same as in Radack. The language is peculiar; the tattooing is different: it extends over the sides of the body, and is spread over the exterior part of the thighs and legs. There are no domestic animals there: the bread-fruit, the cocoa, the banana-roots, and, as at Radack, the fruit of the pandanus serve for their food.

The natives of Repith-Urur live in constant wars among each other. The man always has his arms about him, and, when he sits down to his meals, he lays one lance on his right, and another on his left. Human flesh is eaten at Repith-Urur.

A long time ago, a boat, with five men from Repith-Urur, came on the island of Relick, of the Ralick chain. They fished, but caught nothing There was no want of fruit. They killed one of their number, baked, and ate him. A second was butchered, and devoured in the same manner. The inhabitants of Relick overpowered them, and killed the three that remained.

In the island of Airick, of the Kawen group, there live a man and a woman ; on the group of Arno, two men and one woman, from Repith-Urur, who were cast in boats on Radack. A second woman, whom the latter had with them, died of thirst at sea during their long wandering. These five people were still in Radack on Kadu's arrival there. While he was there, two boats from Repith-Urur arrived at the group of Aur, where he was at the time, in each of which there was a man and woman ; they said they had been nine months at sea, and had lived five months on fish, and without fresh water. The natives of Radack wished to take up arms against these cannibals. The chiefs protected the strangers ; a chief of Tabual received a man and a woman, and a chief of Aur the others.

Bogha is the name of an inconsiderable, low group of islands, with which the Radackers became acquainted by the following circumstance. A woman of Bogha was washed away by the tide, as she was drawing a load of cocoa along the reef from one island of this group to another. Her cocoa served her for a raft, and bore her up; she was driven by the wind and current past Bygar, and, on the fifth day, was thrown upon Udirick. This woman is still living in the island of Tabual, of the group of Aur. Bogha, in its remote situation, appears to us to be the seat of a forgotten colony from Radack, the language of which is spoken there.

The islands seen by Captain Johnstone in the frigate Cornwallis, in the year 1807, in the north-east of Radack, and for which we looked, (the same, according to Krusenstern's Contributions to Hydrography, p. 114, No. 24, p. 19, as seen by Ferdinand Quintana, in the ship Maria, 1796, and the Nassau fleet in 1625, as also the Gaspar-rico of the ancient charts,) form a low sickle-shaped group of inconsiderable circumference, the convex side of which is turned to the leeward. Only on the windward side mould has collected on the reef. It rises mostly naked out of the waves under the lee, and at its entrance sinks into the inner sea. The islands form a very close row; the vegetation appears to be poor; and the cocoa-tree is nowhere seen to rise above the rest.

The desert appearance of this group, and the num ber of sea-birds, and frigate-birds, which swarmed round us near it, and which darted at the red streamer of our ship, as at prey, convinced us that they were really uninhabited; and we cannot agree with our friend Kadu, who wished to recognize Bogha in them. The north-east monsoon, and the strong westerly current which we experienced in the passage from Udirick thither, as it may be constantly expected in these seas, connected with the story of the woman in Tabual, indicate a more easterly situation for the group Bogha. Perhaps it ought to be looked for to the east of the direction pointed out by Udirick and Bygar at a smaller distance from Radack.

That the cocoa trees in Bogha are but low, and the people possess no boats, perhaps passed from the preconceived opinion of our friend, that the islands before him were Bogha, into his description of this group, of which he first began to speak on this occasion.

THE CAROLINE ISLANDS.

The ingenious Pedro Fernandez de Quiros, in 1605, was for looking to the south for the mother of so many islands (*en demanda de la madre de tantas islas*) which had already then been discovered in the Great Ocean. We have found this mother in the continent, to the east of which they are situated, as the sea-fowl are met with to windward of the rocks, which are their native country, whether they return to their nests every evening with the setting sun.

This image, which is particularly applicable to the islands of the first province, forced itself again upon us, as we returned, from the distant Radack, in the east, to the western Carolinas; from the strayed child to the children in the mother's lap. A more bountiful nature receives us, and the same people, with the same amiableness, are more cultivated.

The seas in which the Carolinas are situated, are subject to violent storms, which mostly mark the change of the monsoon. These hurricanes, which the Spaniards on the Philippine and Marian islands call by the Tagalese name Bagyo, sometimes destroy all the fruits on the low islands, so that the inhabitants are compelled to subsist for a long time only by the fishery. They endanger the islands themselves, against which they raise the sea. Kadu, at Mogemug, witnessed a hurricane, during which the sea washed away an island, which was indeed uninhabited, but covered with cocoa-palms and bread-fruit trees.

Captain Wilson gives us a view of the nature of the Pelew islands, and of their productions. Eap, the other high land to the west of the Carolinas, appears to us, though it is without high peaks, as the seat of volcanic powers. Earthquakes are frequent and violent, and even the slightly built habitations of the natives are overthrown by them. When the earth trembles at Ulea, the coral reefs of Mogemug and Ulea are shaken, though with less violence Kadu has not heard that the same happens at Feis According to his observation, the nights are much cooler at Eap than at Ulea, though the days are equally warm. Eap produces whet-stones, which the low easterly islands obtain from thence. They are a kinder gift of nature than the silver, which Cantova, on the testimony of the native, Cayal, ascribes to this island. Kadu has explained to us this tradition. A white stone is found in the mountains of Eap, to which the chiefs have an exclusive right; their seats of honour are made of it. One block forms the seat, and another the back Kadu has seen this stone; it is neither silver nor metal. A yellow stone has the same honour in Pelli (the Pelew islands).

We may remember, in Wilson's voyage, the seat of a chief carried away as a military trophy. A sort of potters' clay is used at Eap as well as Pelli, where they make longish vessels of it. This art cannot exist in the low islands without the material.

The several useful palms of the Philippines (*Palma brava*, *Palma de Cabello negro*,) which are mentioned among the plants of the Pelew islands, allow us to judge of the richness of their Flora. Eap enjoys, with Pelew, the advantage of a high land. We find among the productions of Eap the areca-palm (*Areca catheca*), the bamboo, and three species of trees growing in the mountains, with the wood of which they build their boats, for which the lower islands only use the bread-fruit tree; the *Aleurites triloba*, the clove-tree, (*Caryophyllus aromaticus*,) which is not only not esteemed but despised, and also two other trees, which are useless, and of a bitter taste, and serve as a comparison for what is bad and ugly; the orange-tree, the sugar-cane, and, lastly, the curcuma, which it is true occurs at Ulea and the other low islands, but in greater abundance at Eap. Kadu recognized, in the Sandwich islands, and among the seeds, which were thrown on the reef of Radack, many that were partly native at Eap, and partly in the low islands of the Carolinas. Feis has, of all the islands, the richest soil, and the most luxuriant Flora. The bamboo, which, on account of its manifold uses, was transplanted there from Eap, thrives well. The other groups of islands procure what they want from Eap. Ulea and similar low islands of these seas produce many kinds of plants that are not at Radack, and nature is far more bountiful to them. Don Luis de Torres has even brought plants from Ulea to Guahon, which were foreign to the Flora of that high country

All these islands are rich in bread fruit-trees, roots, and banana. The inhabitants of the low islands seem principally to subsist on the bread-fruit tree, many kinds of which, bearing large fruit, are cultivated under different names. On the high islands, the roots are the chief support of the people. The sweet potatoe (*Camotes* *) which, together with the seeds of other useful plants, Cayal, three of his brothers, and his father, Corr, brought back to Eap from the Bisayas, (Philippine islands,) whither they had been driven, spread from thence to other islands (*see* Cantova); but, according to Kadu, it does not grow at Ulea. The roots of the arum species attain to perfection only on high land, and likewise in Feis. In the Pelew islands† different varieties of one kind are culti-

* The Spaniards call the sweet roots *Camotes*, and they seem to have borrowed this word from the language of the Philippines. The *Camote* of the Tagalese and Bisayas was cultivated on these islands before the conquest.

† In the account of the Pelew Islands, there is every where yams, *i. e. Dioscorea*, by mistake, for taro, or arum. *Lin.*

vated, some of which attain to an extraordinary size The pandanus grows on all the Carolinas, where, however, the fruit is neither eaten, nor even used for an ornament. None of the improved kinds are found there. The agriculture of Eap must be admirable. Swimming plantations of arum are there ingeniously constructed on rafts of bamboo and wood on the water.

The Plantain is not cultivated so much for its fruit as for its fibres, of which the women weave or braid a pretty stuff, resembling mats. The pieces of this stuff are in the shape of Turkish shawls, three quarters of a yard broad, and several yards long. Black threads woven between, form an elegant pattern on both ends, and these threads hang down as fringes. This stuff is sometimes dyed with curcuma. It is described, in the Voyage of Captain James Wilson, in the Duff, who, in 1797, had intercourse with the islanders of the province of Ulea, where the art of making it is ascribed, without any reason, to the instruction of the Spanish missionaries.* According to Kadu, the banana-plants are mostly cut down before they bear fruit, for the sake of the fibres.

Another plant, a *Malvacea*, furnishes a bast,

* We can easily conceive, that the inhabitants ask for iron by the name by which Luito, nine years before, obtained much from the Europeans at Guahon (*Lutu*, Chamori, for Parang, Ulea). But we do not comprehend how it happens, that the numbers given are of no dialect of this sea known to us. We recognize only the general roots of the language.

which is likewise used in some islands for similar stuffs.*

The paper-mulberry-tree, and the bast-stuffs of Owhyee, were equally unknown to Kadu.† The curcuma-root, which is rasped to a powder, forms a considerable branch of the trade of Eap. The custom of dyeing the skin with this powder, is general, from Tuch in the east, to Pelli in the west, but is not prevalent in the groups to the south-west of the Pelew islands, neither did it prevail in the Mariana islands. The women always ornament themselves in this manner, and the men only on holidays, or in war, for battle; and thus too the bodies of the dead are adorned. The custom of chewing the betel, and of dying the teeth black, is exclusively confined to Pelli, Ngoli, Eap, and the Mariana islands, where it formerly also prevailed. It is only in the Pelew islands that a sweet syrup is obtained from the cocoa-palm. The drinking of kava, and the use of salt, are equally unknown in these islands

* A passage in Cantova's letter confirms us in the conjecture, that the sterile banana, which is only cultivated in the Philippines on account of the flax, also grows in the Carolinas. "Mettre en oeuvre une espèce de *Plane* sauvage et un autre arbre qui s'appelle *Balibago* pour en faire de la toile."

† A passage in Pigafetta might lead us to conjecture, that the little apron which the women wore in the Mariana islands, was of stuff made of bast. "Toile ou plutôt écorce mince comme du papier que l'on tire de l'aubier du palmier."—*Page* 61. *of the French edition.*

No domestic animals are to be found on the islands of the first province of the Great Ocean, except those which are brought by Europeans. We leave it to Wilson to speak of the Pelew islands. According to Kadu, a very considerable time ago, a large ship had come to Mogemug, which had left some cats there. The species of this animal has extended from Mogemug to the west as far as Pelli, and to Ulea in the east. They are called in these islands by the Spanish name *Gato.* The people of Eap and Ulea, and Kadu himself, learnt, from a very old man at Mogemug, to count from one to ten in the language of these strangers. So far, in fact, he counts in the Spanish language with fluency, and correct pronunciation. He has likewise seen, at Mogemug, two large earthen vessels, (three or four feet high), which were left there by that ship. We have traced no other memorial of the mission of Cantova to Mogemug. Kadu heard nothing of the cannon left at the island of Falalep.*

The *Trichechus dugong* is found in the waters of the Pelew islands, as in those of the Philippine islands.

Cantova mentions the chase of the whale by the inhabitants of the low islands. Perhaps what he relates might be understood of the dolphin fishery.

* Caschattel, lord of Mogemug at the time of the letters of Cantova, was known by name to Kadu, as a long deceased chief of this group.

There are three kinds of dolphins in these seas, with a white, red, and black belly As soon as any of these animals are observed from Ulea, small boats, to about the number of eighty, put out to sea, surround the herd, drive it on the shore, and, when they have approached it near enough, they pelt the animals with stones till they have thrown themselves on shore. In this manner they catch great numbers of them Their flesh is considered a great delicacy. In the cutting up particular rules have to be observed. A wrong cut causes the animal to keep from the island for a certain time. At Svilick, where the reef has only a narrow entrance, these animals are driven into the lagoon, and none of them are killed until a certain number (about fifty) are caught. This chase is executed on the islands belonging to Ulea, with particular success. In the others the art is not so well understood. The dolphins sometimes ascend the rivers of Eap, when the inhabitants intercept their return with nets, and harpoon them.*

The domestic fowl is met with on all the Caroline islands, though the inhabitants do not understand how to derive any particular advantage from it. Against the account of Cantova, who reports to us what he heard himself from the natives of Eap, that a kind of crocodile is worshipped, or

* The inhabitants of Eap have large nets for fishing, which are not, and perhaps could not, be used on the low islands.

honoured there, we must quote expressly the testimony of Kadu.

In Pelli (the Pelew islands) is a sort of crocodile called *Ga-ut*, (*Ye-use* according to Wilson.) The *Ga-ut* lives always in the water, and has a flat tail. The sound, resembling the voices of children, which, this dangerous animal utters, may deceive persons who are ignorant of it. The *Ga-ut* of Pelli is not found at Eap. It was only once that one of them was seen there, and was killed, after it had devoured a woman.

A large species of lizard, called *Kaluv*, is found on the islands of Pelli and Eap exclusively, and is expressly stated not to be found at Feis. The *Kaluv* is much smaller than the *Ga-ut*, and its tail is round. It goes indeed into the water, where it may be dangerous to man, and eats fish, but it lives more frequently on shore, and crawls on the trees, where it sleeps during the day. Kadu recognized the *Kaluv* in the figure of the *Lacerta Monitor*, which Sonnini and Latreille give in the "Suites à Buffon;" the flesh of this animal is considered as poisonous at Eap, and is not eaten. The natives think it would occasion death, but they kill the animal whenever they can. Boclé, the adopted son of the chief and priest of the territory of Kattepar, and his companions (perhaps Europeans) ate the flesh without fear and without bad consequences.

Among the insects of Eap which are not found in other islands, Kadu mentions a very large scorpion, the sting of which, said to be mortal, is cured by the juice of herbs; and a small species of *Lampyris*, which is only met with in some districts. The flea was entirely unknown to Kadu till he came to us.

Iron of wrecked ships, which are thrown on shore at Ulea, Eap, and other islands, is more abundant than at Radack. None is said to be found to the south-west of Pelli. Drift wood is every where neglected.

Cantova mentions a mixture of different races of men in the Carolinas, respecting which our accounts are silent. It is possible that some Papuas of the southern countries may, by accident, have reached these islands, and some Europeans, Martin Lopez and his companions, or others, in different ways, as has since frequently been the case. But the race of the natives, is that which has spread over all the islands of the Great Ocean. Their hair seems to be more curly than that of the inhabitants of Radack. They let it grow long, and place a particular value on this natural ornament. It is only at Eap that that of the children is cut.

According to Kadu's observations, the inhabitants of the territory Summagi, in Eap, are of remarkably short stature. Deformities and natural defects are uncommonly frequent in that island. He once mentioned, as an example, a man without arms, whose head was uncommonly large; another without hands; another without thumbs; a person with only one leg, instances of hare lips, and of persons deaf and dumb.* Even less remarkable cases are much more rare in other islands. A sickness, which the Europeans have spread in most of the islands of the South Sea, seems, according to Kadu, not to be quite unknown at Ulea.

The inhabitants of the Carolinas in general are fatter and stouter than those of Radack. Tattooing is every where arbitrary, and has no reference to religious faith: the chiefs are not more tattooed than the common people. A piece of banana stuff, worn almost like the maro of Owhyee and Otaheite, is the usual dress, and only at Pelli the men are entirely naked, as was also formerly the case in the Mariana islands. The ear ornaments of the inhabitants of Radack are not worn at Pelli. The gristle of the nose is pierced through for putting sweet-scented flowers in. The bracelet, made of the bones of the *Trichecus dugong*, which the chiefs of the Pelew islands wear, is described by Captain Wilson. The chiefs at Eap wear a similar broad bracelet made of a shell polished.

The houses are every where large, and closed. You can come in at the door without stooping. Paved foot-paths, and square yards are before the houses at Eap, as in the Pelew islands, where we have been made acquainted with them by Captain Wilson.

* Kadu also saw at Eap a monstrous *Kaluv*, that had two tails, and two tongues.

We must first of all contemplate this courageous seafaring people in their boats.

The boats of Nugor and Tuch, the inhabitants of which are distinguished by a distinct language, and those of the lower islands as far as Ulea, Feis, and Mogemug, which speak the same language, are, according to Kadu, of the same construction as the boats of Ulea. The inhabitants of Savonnemusoch, between Nugor and Tuch, who speak another language, do not undertake long voyages, and probably have other boats. The comparison which Cantova has made between the boats of the Carolinas and those of the Marianas, enables us to judge what these were. The boats of the Marianas were similar to those of Ulea, which are, however, preferable, and better sailers.*

* The two boats which Cantova observed, were overtaken by the west wind, with four others, when on their voyage from Fatoilep to Ulea, and dispersed. Most of the people were natives of those two groups, and we take it for granted that their boats were of those islands. The first large boat, in which were four-and-twenty men, had three cabins, and on account of its remarkableness was particularly described; namely, "Une barque étrangère peu différente des barques Marianoises, mais plus haute;" the other, smaller one, "Une barque étrangère quoique semblable à celle des îles Marianes." He says further, when the distance of the islands from each other is to be estimated, "J'ai fait attention à la construction de leurs barques qui n'ont pas la légereté de celles des Marianes;" and we believe that we have shown, in another place, where there was no other standard, that the distances were still assumed too great. Ulea is even placed at a less distance from Guahon, probably on account of the false determination of the situation of Fatoilep, by Jean Rodriguez. in 1696, on whom Cantova relied.

The construction of the boats at Eap and Ngoli, differs but little from that of Ulea. But the natives of Eap are very fond of using the boats of Ulea, which they procure in the way of trade. Pelli has a construction of its own; and the low islands, to the S.E. of Pelli, have, again, another. Pelli, and these islands, are inferior in the art of navigation; and their boats do not visit the eastern islands.

The natives of Ulea, and of the islands in its vicinity, are the most skilful navigators; and Cantova also considers them as being more civilized than the others. * The main-spring of navigation is trade. The principal articles of trade are iron, boats, stuffs, and Curcuma powder. We have spoken, in another place, of the trade with Guahon, where the people of Ulea principally sell boats for iron. Those of Feis, Eap, and Mogemug go to Ulea to purchase boats, in exchange for Curcuma powder. Those of the eastern islands, which have bread-fruit trees in abundance, build all their boats themselves; and those of Nugor and Tuch barter stuffs at Ulea for iron. Those of Ulea go to Tuch and Nugor; and those of Savonnemusoch are visited in these voyages, but do not visit any other islands. In Pelli, the iron which the Europeans bring there is exchanged for Curcuma. On the south-western groups of islands they exchange stuffs

* Les habitants de l'îsle d'Ulea et des îsles voisines m'ont paru plus civilisés et plus raisonnables que les autres.

for iron, of which they are in want. A squadron of ten sail, five from Mogemug, and five from Eap, performed this voyage; and Kadu was personally acquainted with the seamen at Eap.

They are guided in their voyages by the starry firmament, which they divide into constellations, giving each its particular name. *

They seem to observe, in every voyage, the rising or setting of another constellation. A misunderstood expression of Cantova has caused the knowledge of the magnetic needle to be ascribed to them. † Cantova means only the division of the horizon into twelve points, which we have given with other denominations of the rhumbs and winds, in our vocabulary, after Don Luis de Torres, and Kadu. The steersman of a boat lays, according to D. Luis, a piece of wood, a little stick, flat before him, and considers himself guided by this, as we by the compass. It seems probable to us, that this stick, placed in the moment of observation, in the seas, where the winds are constant, may serve to represent to them the way they have to keep, with respect to the wind.

They count the days and months on the Caroline islands, and divide the year into seasons, according to the disappearing and re-appearing of the constellations; but nobody counts the years. What

* According to Cantova they learn astronomy: "Le maitre a une sphère, où sont tracés les principaux astres."

† Ils se servent d'une boussole qui a douze aires de vent.

is past, is past. The song records the names which are worth preservation, and the multitude float carelessly with the stream. *

Kadu knew as little about his own age, as any of the islanders of the Eastern Polynesia. The life of these islanders, thoughtless, resolute, and only belonging to the moment, are free from many of the torments which undermine ours. When we told Kadu of the suicides among us, he thought he had heard wrong, and it appeared to him the most ridiculous thing we had ever told him. But they are, and for the same reason, impatient under foreign deliberate oppression; and history records in her annals, the suicide of the Marians, under the Spaniards, (the messengers of the Gospel?)

In all the Caroline islands, only invisible, heavenly divinities are believed. Nowhere do they make the figures of gods; nowhere adore man's works, or corporeal things. Kadu knew very little of the theosophy of his people. What we here relate after him, leaves many things to be wished for, and, perhaps, needs correction. We have thought, that we must translate, by the word God, his word Tautup, (*Tahutup*, C.) and Jageach, at Radack. According to Cantova, the Tahutup are departed souls, which are considered as guardian genii.

* "Carpe dien."

The god (*Tautup*) of Ulea, Mogemug, Eap, and Ngoli, is called *Engalap;* of Feis, *Rongala;* of Elath and Lamureck, *Fuss;* and of the desert island, Fajo, *Lagé.*

Is Engalap, the Eliulep of Cantova, Aluelap of Don Luis de Torres, the great God?

No man ever saw Engalap. The fathers have delivered the knowledge of him to their children. He visits distant islands where he is acknowledged. The time of his presence seems to be the season of fertility. He is united to Rongala, god of Feis, by ties of friendship; and they hospitably visit each other. He seems to have no connection with Fuss, the god of Lamureck. At Ulea, and the more easterly islands, (*Lamureck,* &c.) there are neither temples nor priests; and no sacred offerings take place. There are several temples built at Mogemug, Eap, and Ngoli; offerings are made there, and they have a religious worship.

Kadu has informed us, how it was at Eap, where he had lived for a considerable time, and affirms that it is the same in the two next groups. Both sexes have different temples, and different times for offering. No man is present at the offering of the women. At the offerings of the men, the chief acts the part of the offering-priest. He consecrates to the god a fruit of each kind, and a fish, by holding it up, and by invocation. The formula is, *Wareganam gure Tautup,* the people repeating the ast word. The offered fruits are not eaten, but deposited in the temple. To prepare for these offerings, the people remain assembled, and without communication, for a month in the temple, where their food is handed to them from without. Every one consecrates the first piece of all the fruit and fish which he eats during the time, in the above manner, and then throws it away, without having tasted it. Neither songs nor dances take place in the temple. This solemnity is alternately for one month in one territory, and on the following in another. Kadu, being a stranger, did not attend the solemnities in the temples. He never went into them. Access to the temple is forbidden to every one, out of the time of offering, except to the chief or priest, (*Matamat.*)

Rongala has no temples at Feis. But there are times at which he descends on the island, and is invisibly present in the wood. At that time, the people are not allowed to speak or tread loud; they then approach the forest, dyed with Curcuma, and adorned as for a festival.

We will communicate faithfully, and in detail, the mythology of Ulea, according to Don Luis de Torres. Cantova, whom we beg to be allowed to compare here, relates the descent of the gods, almost in the same manner, and a little more circumstantially. The pleasing fable of Olifat is entirely new.

Three persons are worshipped in heaven, Aluelap,

Lugeleng, and Olifat. The origin of all things is, however, as follows: before all time, there was a goddess, called Ligopup. She is regarded as the creator of the world. * She bore Aluelap, the lord of all knowledge, the lord of glory, the father of Lugeleng. † But it is not known who was Lugeleng's mother, or any thing respecting his birth. Lugeleng had two wives, the one in heaven, the other on earth. The heavenly one was named Hamulul, the earthly one Tarisso, who equalled the first in beauty, and other natural gifts.

Tarisso bore Olifat ‡ from her brains, after a four days' pregnancy. Olifat ran away as soon as he was born, and he was followed to cleanse him from the blood: but he said, he would rather do it himself, and would not allow any body to touch him. He cleansed himself on the trunks of the palm trees, past which he ran; for which reason they have retained their reddish colour. They called to him, and followed him, to cut off the navel-

* According to Cantova, Ligopund, sister and not mother to Eliulep (Aluelap, T.) creatress of man. The first of gods are, however, Sabucur, and his wife Halmelul, the parents of Eliulep and Ligopund.

† Lugueileng, according to Cantova, who calls his mother Leteuhieul, born at Ulea.

‡ Oulefat, Cant.—He does not name the wives of Lugueileng, but makes the terrestrial mother of Oulefat to be born in the island of Falalu, in the province of Hogoleu. This island is not known to Kadu; it is called Felalu on the chart of Don Luis de Torres.

string, but he bit it off himself; he said he would take care of himself, and would not suffer a mortal to touch him. He recollected that it was customary to give to new-born infants the milk of a young cocoa-nut to drink, and came to his mother, who gave him a cocoa to drink. He drank, and turned his eyes towards heaven, where he perceived his father Lugelang, who called to him. He then followed the call of his father, with his mother. Both, therefore, left the world. When Olifat arrived in heaven, he met there several children, who were playing with a shark, round the tail of which they had tied a string. To remain unknown, he pretended to be leprous. The children thereupon kept at a distance, and would not touch him. He asked them for the fish to play with, but they refused him. One of them, however, had compassion and gave him the string to which the fish was fastened. He played with it for a short time, and then returned it to the children, desiring them not to be afraid, but to continue playing, the fish would do them no harm. He, however, bit all of them, except the one who had been kind to Olifat. Olifat had cursed the shark, which before had no teeth, and was harmless. He then went through heaven cursing all creatures, on similar occasions, because he was provoked in the glory. As nobody knew him, and he had not yet come to his father, who was alone able to recognize him, his life was sought to be taken. He came to a place where a large house was building; he asked the workmen for a knife, to assist in cutting cocoa-leaves for the roof, but they refused him ; one, however, gave it him and he cut a bundle of leaves, but he cursed all the workmen except the one who had favoured him, so that they were changed into immoveable statues. Lugelen, however, the master of the house, inquired for his workmen, and he was told that they had been turned into immoveable statues. By this Lugeleng and Aluelap knew that Olifat was wandering about in heaven. They asked the man who was still employed in cutting cocoa-leaves for the house, whether he had not seen any thing about, and he answered, that he had seen nothing but a *Canduru*, (a kind of sand-snipe, into which bird Olifat had changed himself.) They sent the man to call the *Canduru*, but when he did it the bird was frightened at the voice, and flew away. The man reported it, and the gods asked him what he had told the bird. He answered, he had told him to come. They sent him again, and commanded him to order the bird to depart, because he obstructed the chiefs. He immediately did so, and the bird instantly came. He forbid him to go farther, and to sit in the presence of the chiefs, and the bird immediately did what he was forbidden. As soon as he was seated, Lugeleng commanded the workmen who had been petrefied in the wood to be called together, and they soon came, to the great astonishment of the spectators, for Lugeleng and Aluelap alone knew that it was Olifat.

The workmen now continued with the building, and dug deep holes in the ground to erect the posts. This appeared a fortunate opportunity for those who intended to kill Olifat, on account of the much mischief which he caused. Olifat, however, was acquainted with their intention, and concealed about him some dyed earth, coals, and the rib of a palm-leaf. He now digged in the hole, and made a side opening to hide himself in it. But they imagined that it was now time, threw the post in, and mould round its foot, and thought to squeeze him to death. But he took refuge in the side opening, spit out the dyed earth, and they thought it was his blood: he spit out the coals, and they thought it was his gall. They now imagined that he was dead. With the assistance of the rib of the cacoa-leaf, Olifat made his way through the middle of the post and escaped. He laid himself as a beam directly across the post from which he had come, and was not perceived. When the day's work was over, the workmen sat themselves down to their meal. Olifat sent an ant to fetch him a piece of cocoa: the ant brought him as much as it was able. By his power he enlarged it to a whole nut. He then cried aloud: "Take care below, I am going to break my cocoanut." They recognized him by this exclamation, and were much astonished that he was alive. They considered him to be Alus, the evil spirit. *

* Nombre que dan al Diablo.

They still resolved to kill him, and told him that he should only finish his meal; they would afterwards give him a commission. They sent him to the house of Thunder to bring him his meal. Olifat took a reed and went away. He came to the house of thunder, and said imperiously and rudely, "I have tired myself by bringing food for your misshapen mouth." He gave him his meal, and went away. The thunder intended to fall upon him, but he concealed himself in his reed. The thunder could not find him, and gave up his search. Olifat came again forward, and created the greater astonishment, as he came back after this trial without any hurt. The workmen again sent him out to bring the meal to the fish *Fela*.* Olifat entered the house of the fish *Fela*, but as he was not at home himself, he threw the meal to those who were there, saying, "Take it for yourselves," and departed. When the fish came home, he asked for him who brought the meal. The family told him, that somebody had thrown the food to them, but they did not know who he was, nor whither he had gone. The fish now began to throw a hook, fastened to a long line, to all the winds; and when he at last threw it to the north, he pulled out Olifat, and killed him. After a lapse of four or five days, without Olifat appearing, those who sought to kill him in heaven, consoled themselves,

* This is the fish whose upper jaw is so much shorter than the under one.

and thought he was now dead. But Lugeleng looked for his son, and at last found him dead, and full of worms. He raised him up in his arms, and awakened him. He asked him who had killed him. Olifat answered, he had not been dead, but had been sleeping. Lugeleng called the fish *Fela* to him, beat him with a club, on the head, and broke the upper jaw: hence the form which he now has. Aluelap, Lugeleng, and Olifat, now entered into glory, where they are engaged in administering justice.

Others fix the number of divinities at seven, namely, Ligopup, Hautal, Aluelap, Litefeo, Hulaguf, Lugeleng, and Olifat.

To the question whether other islands had the same belief, several answered, That this was the belief of the whole world, and the world would perish when Aluelap commanded it.

For comparison we shall communicate the doctrine of the former inhabitants of the Mariana islands. Velarde, tom. ii. fol. 291. Puntan was a very wise man, who had lived many years in the void space, before the creation of the heaven and the earth. When he was about to die, he commissioned his sister that she should make the heavens and earth out of his breast and shoulders, the sun and moon out of his eyes, and the rainbow out of his eye-brows. *

* Thus in our northern mythology:—

Or Ymis holdi	Literal:	Out of Ymer's flesh,
Variörth om scavpvth		Was the earth created,

Though there is no public service at Ulea to the gods or the divinity, we were yet assured by Don Luis, that the people were not without pious sentiments. Individuals sometimes lay fruits, as offerings to the invisible beings, and nobody is blamed for taking this offering and eating it.

Cantova mentions a peculiar manner of enquiring into one's future destiny. The mode of proceeding is the following: They tear two strips from each side of the rib of a cocoa-leaf, repeating the syllable, *Pué-pué-pué*, one after the other, then hastily make knots in every strip, repeating the question they intend to put respecting their fate, in intelligible words. The first strip is laid between the little and the ring finger, with four knots in the inside of the hand, and the other with a decreasing number of knots between the middle and the fore-finger, and between the fore-finger and the thumb. According as the number of knots hanging over the back of the hand coincides with, or differs from the number of fingers, one, two, three and four, the indication is fortunate or unfortunate.

At Ulea, as among other people, many customs,

Enn or beinam biörg,	But out of (his) bones rocks;
Himinn or havsi	The heaven out of the skull
Ins hrimkalda iotvnns.	Of the ice-cold giant;
Enn or sveita siòr.	But out of his blood the sea.

Vafthrusdismal XXI. Edda saemundar, p. 13.

depending on their belief, are observed, and many conjurations prevail. We have mentioned the cutting of the dolphin. A small fish is caught very often, with which children are not allowed to play. Should it happen that any person held this fish by the tail, so that the head hung down, all the fish would, in the same manner, seek the bottom of the sea with their heads at the next fishery, and none would be caught. Many people are not allowed to eat of the fruits of the banana bunch. Only that person who has eaten one of the bananas is suffered to eat the rest.

On the desert island of Fajo, as at Bygar, fresh water is conjured into the water pits.

There is a species of black-bird, which is under divine protection on this island, and not permitted to be eaten.

The people of Eap are famed for their skill in conjuration. They understand how to conjure the sea, to lay the storm, and, in a calm, to call up the wind from the favourable quarter. They understand, by throwing, with conjuration, an herb into the sea, to excite the waves, and to raise the most dreadful storms. To them is ascribed the destruction of so many boats from Feis and Mogemug, nay, even the gradual depopulation of these islands. In a piece of fresh water, in the territory of Sutemil, there are two fish, only a span long, but extremely old; they keep constantly in one line, with their heads turned towards each other. If one of them is touched with a switch, so that it moves forward, and they cross each other, the islands are shaken to their foundations, and the earthquake does not cease till both occupy their former situation. Above these fish, and the water in which they live, a house is built, and over it watch the chiefs, at whose death there sometimes occurs an earthquake.

A certain Eonopei (he is now dead, his son Tamanagack is a chief of the territory of Eleal) once showed our friend Kadu a remarkable specimen of his art. Eonopei prepared a round flat cake out of taro dough. It was night and full moon. He began, with conjurations, to eat his cake. In proportion as he bit into the edge of it, the full disk of the moon was impaired and hollowed out, more and more, in the form of a sickle. When he had thus, for a time, magically eaten at the moon, he altered this process and his conjurations. He now began to knead the soft dough of his cake, which remained, again into a full circle; at the same time the disk of the moon equally increased, and at last the moon shone quite full. Kadu, during this time, sat next to the conjuror, looked at every thing, the moon and the cake, with the greatest attention, and wondered how the circle of each could, simultaneously, first decrease and then encrease. We make no comments upon the account of our simple-minded friend, of whose veracity we have not the least suspicion, and leave it for more learned commentators, to apply it to an eclipse of the moon; which we cannot, however, suppose could be foretold at Eap before the discovery of writing.

Feasts and banquets, which take place on several occasions, such as the piercing the ears of children, the cutting of their hair at Eap, tattooing, &c. seem to have no religious signification.

Singing and dancing, mostly inseparable, every where constitute the chief delight, the chief amusement. There are different kinds of festive sports, which are led up by the different sexes, or by both united, and of which each bears a different character and a peculiar name. Their songs are, however, not accompanied by any musical instruments, and even the drum is unknown in the Carolinas.

The chiefs seem to be subject to each other, according to a kind of feudal system. Opinion raises them high above the lower people, and extraordinary demonstrations of respect are paid to them, with which we are sufficiently acquainted by Cantova's Letters, and (for Pelew) from the Account of the Pelew Islands. They bow before them to the ground, and only crawl to them. The boats lower their sails in sight of the island of Mogemug, the residence of the principal chief of the group of this name. This reverence to noble, perhaps divine, descent, does not appear to interfere in purely human connections between chiefs and other men, notwithstanding the relative rank, to which due regard is paid. The principal chiefs have great authority, and exercise penal justice, according to the principles of the strictest retaliation; "An eye for an eye, a tooth for a tooth."

According to Cantova, criminals are only punished by banishment. We will relate a story, according to our friend Kadu, in which it is evident with what great lenity crimes are less sought to be revenged than repressed. We fancy we hear the national tale of the *Fin-voleur* from the mouths of our nurses.

On an island of Mogemug the trees were regularly robbed of their best fruit, without the people, who carefully watched each other, being able for a long time to discover the thief. They were at last aware, that an apparently well-disposed boy got up every night and committed the robbery. They chastised him, and watched him. He however deceived their vigilance, and did not leave off his practice. They shut him up during the night; they tied his hands on his back, but the sly thief knew how to frustrate all these precautions, and continued as before. They brought him to a remote, uninhabited island of the group, which barely afforded food for one man. Here they left him alone. They, however, soon perceived that it had been of no avail, and their trees were still robbed as before. Several persons rowed over to the desert island, and found the young man feasting in great abundance on the fruits of their property. The trunk of a tree served him for a boat, and he went out every night on his depredations. They destroyed this vessel, and left him, having deprived him of the power of doing mischief in his solitude. They had now rest. After some time, they wished to know how he did, and some persons went again to the island. They heard and saw nothing of him. After they had in vain called and looked for him in the woods, they returned to the shore, and found that their boat was gone. The sly thief had gone with it to sea. He sailed over to Sorol. On this group he did not leave off his tricks, but projected greater undertakings. He persuaded the chief of Sorol to make an attack on Mogemug. He was to surprize the chief in the night, and murder him, and seize upon the supreme power. The confederates came by day in sight of Mogemug. They lowered their sails, awaiting the night in the open sea. The boat was, however, observed, and they were surrounded on their landing. The instigator was punished with death. The people of Sorol returned uninjured to their island.

The succession at Ulea, as at Radack, descends first to the brothers, and then to the sons of the eldest son.

According to Kadu, the chiefs give to the eldest son the name of their father, to the second the name of their wife's father, to the third again the name of their father, and so on; other people, on the contrary, give to their eldest son the name of the father of their wife, and to the other children different names; and it is said to be the same at Radack. According to Don Luis, the degree of consanguinity is indicated in the names, and it is to be recognized by them whose son or nephew he is.

The amicable exchange of names, a general custom in Eastern Polynesia, is unknown on the Carolinas, and Kadu at first denied that it was customary at Radack, though he himself gave us examples of it in the sequel.

Marriages are concluded without any solemnities. The man makes a present to the father of the girl whom he marries, consisting of fruits, fish, and similar things. The value of this gift is according to the rank of the father of the bride, for marriages also take place between persons not equals. If only the father, or only the mother, is of the class of the chiefs, the children are accounted to belong to this class. In the first case the man and husband shows to his wife, and his children by her, all the exterior demonstrations of respect due to their rank. A plurality of wives is allowed. Marriages are dissolved, as they were concluded, without any ceremony. The husband sends his wife back to her father. The husbands also cohabit with their wives when they are pregnant, but not when they have a child at the breast. The latter is only customary at Radack; the former, contrary to Wil son's assertion, is expressly observed at Pelew. There a chief, who has generally several wives, has his place, with any one of his wives who is in this state, supplied by a chosen man. We shall speak particularly of the customs of Pelew. Married women in other islands are entirely devoted to their husbands. They are bound by duty, and the uncorrupted morals of the people seem to guard their virtue. Unmarried women are allowed their liberty, according to custom. They pass the nights in their own large houses. Infanticide is unheard-of; the prince would have the unnatural mother punished with death.

What we related respecting the burial of the dead at Radack, is also customary at Ulea, and the islands lying more east. According to Kadu, the corpses of all persons, without distinction of birth, are buried on the islands at Feis, Mogemug, and Eap. We, however, see at Mogemug, after the great tragedy which concludes the history of the Carolina missions, the customs of Ulea observed towards the dreaded strangers whom they had murdered, and must believe that Kadu erred with respect to Mogemug. At Eap the burying-places are in the mountains. The mountaineers fetch the corpses of persons who have died in the valley, and for this office receive a present of fruits, roots, &c. It appears that none of the friends follow it to the grave.

A bond of inviolable friendship is concluded in

all these islands, exclusively between two men, which binds the friends together with particular force. The chief and the common man can also conclude such a bond, without infringing on the relative rank of the parties. Though this kind of friendship is found on all the islands, yet different duties and privileges are attached to it at different places. At Eap, on every kind of engagement, one friend must stand security for his friend, and where he is injured, or killed, he is obliged to revenge it. To similar obligations a new one is found at Ulea. When the friend claims the hospitality of his friend, he gives him his wife during the time of his visit, which is not done at Feis, and farther west. We have seen, that at Radack the duty in the first respect is less binding; in others, the same as at Ulea.

The touching of the noses, as on the islands of Eastern Polynesia, is the customary salutation.

Of the Carolinas, only Pelew, Eap, Tuch, and the remoter islands, with which Tuch is at variance, are acquainted with war. The other islands, as Ulea, enjoy an uninterrupted peace. Our kindhearted companion frequently and joyfully repeated, "There, there they know nothing of war and combat! there man does not kill man! and he who sees war, his hair will turn white!" War has not always prevailed at Eap. Formerly, the island recognized the authority of a principal chief, and there was peace. But since the death of Gurr, the last supreme chief, the chiefs of the different districts often settle their disputes by arms. In case of any wrong, or offence, the triton's horn is blown. Both parties approach each other armed. They negociate. When satisfaction is refused, and no compromise can be agreed upon, they fight. The war lasts till one of the class of chiefs has fallen on each side, and the opposite party eaten a part of his flesh. Every one only just lifts a piece to his lips. This is an indispensable ceremony. As soon as this condition has been complied with peace is restored; and it is confirmed by marriages between the two parties. The character of these islanders is, however, mild and hospitable as those of the other groups. The stranger at Eap and Pelew goes without danger between the belligerent parties, and enjoys an equally friendly reception among both. The people of Eap throw the dart in an arc, with the assistance of a hollowed piece of bamboo, in which the unarmed end of the weapon is held and receives the impulse. By this means they throw it an extraordinary distance. This weapon seems to agree in the essential parts with those of the Aleutians and northern Esquimaux. They have also the two-pointed lance of the Radackers. When the combatants have approached each other, this lance is thrown straight, and merely by the hand. They at last fight man to man. The chief leads on to war with the triton's horn. The military expedition approaches towards the hostile territory in boats and rafts of bamboo. The landing is attempted to be opposed. The decisive combats take place on shore.

The people of Tuch use, for close combat, the dart, and, at a distance, the sling. Their throw is far and sure; they manage this weapon with astonishing skill. They also constantly wear them in time of peace tied round the head, and use them to kill birds, to beat down fruit from the trees, and the like. Kadu learned to use the sling at Ulea from the natives of Tuch, and he often passed his time among us in this exercise, in which he was, hewever, very unskilful.

Don Luis de Torres extolled, in his friends of Ulea, what gave us pleasure to praise in our friends at Radack: they are kind, friendly, elegant, and modest. A woman never came on board the Maria. They are good-natured, affectionate, generous, and grateful. They have the memory of the heart. Any thing, a useful instrument, for example, which they have received as a gift from a friend, retains and bears among them as a lasting memorial, the name of the friend who bestowed it. Thus Kadu intended to give our names to the animals and plants we had introduced there, to keep us in eternal remembrance.

Of the natives of the Pelew islands, (*Palaos*

Panlog,) Cantova gives a terrible picture.* According to the accounts he received, they are hostile cannibals. They appear to us, in the account of the grateful Henry Wilson, who owed his return to his country to their generous hospitality, in the most favourable light, the partial colouring of affection, adorned with every virtue; and the fact proves, that they exercised most of these virtues. We live with Wilson among this people, see with our own eyes, and judge ourselves. Since Wilson's time, English, Spaniards, and Americans have continually visited the Pelew islands; several Europeans have settled there, and the *Trepang* is constantly collected on their reefs for the mart of Canton. Kadu, who had been at the Pelew islands, gave us a comparison between the two people. This comparison, according to the opinion of our friend, was not favourable to the natives of the Pelew islands. He particularly censures them for total want of modesty, so that they brutally follow the impulse of nature in open day. He excited in

* Peuple nombreux, mais inhumain et barbare; les hommes et les femmes y sont entièrement nues et se repaissent de chair humaine, les Indiens des Carolines regardent cette nation avec horreur, comme l'ennemie du genre humain et avec laqu'elle il est dangereux d'avoir le moindre commerce. Ce rapport me paroit fidèle et très conforme à ce que nous en a appris le P. Bernard Messia, comme on le peut voir dans sa relation." This account is nowhere found, and appears never to have been printed.

us the picture of excessive licentiousness, such as is found in the Sandwich islands.

A few leaves of a journal, communicated to us at Cavite, by a Spaniard, who had passed nine months in the Pelew islands, are drawn up in the language of abuse, not of criticism. He makes less impression on us than our honest friend, whose inculpations he, however, circumstantially repeats. "The man knows the woman in the sight of every body. All are ready to give up their wives for a trifle," &c.; but he also accuses them of eating human flesh, and scarcely allows them the figures of men. We lay down his melancholy writing after having merely mentioned it: they are probably no longer the innocent and unsuspecting friends of Wilson. What they have learnt from us has not made them better.

THE PENRHYN ISLANDS. *

The high, thick woods, which are formed by the cocoa-palms on the Penrhyn islands, deceived us at a distance, with the appearance of elevated shores. Smoke announced to us that they were inhabited. As soon as we had approached the shore, innumerable boats surrounded us, and a peaceable people offered to trade with us.

The islanders are strong and well made, stouter than the inhabitants of Easter Island, and of the same colour. They are not tattooed, but many have furrows and stripes lacerated in the skin of the body and arms, which in one of them seemed to be quite fresh and bleeding. They have frequently no front teeth. Elderly people are very corpulent, and large. We observed several old men who had suffered the thumb-nails to grow, a speaking testimony of their privileged idleness. In one of them, this nail, which was bent inwards, had reached to the length of between two and three inches.

We-counted about thirty-six boats; in each of them were from seven to thirteen people, who

* Voyage of Govenor Philipp, Second Edition, Lond. 1790, p. 233. Lieut. Watts' Narrative of the Return of the Lady Penrhyn, (Capt. Sever,) p. 254, and Appendix, p. 33. table 7, p. 39.

seemed all to belong to the same family. An old man (perhaps the father of the family?) stood in the middle and spoke for all. He had tied, apparently as an emblem of peace, the end of a cocoa-leaf round his neck. There were women in three boats only. In these an elderly woman (the mother of the family?) occupied a back seat, and seemed to have an important influence in the affairs of the men. The authority of no individual seemed to extend farther than his own boat.

The women wear a girdle, with stripes of bast hanging loosely, similar to that of the men at Radack. And the men only bunches of cocoa-leaves tied to a cord. Only a few of them had a scanty covering for their shoulder. This consists of a coarse mat in two pieces, made of a cocoa-leaf. A part of the middle rib, on which the little leaves grow, forms the under-edge of this basket-like mantle. Sometimes bleached pandanus leaves are braided between, for ornament. Only a few of them had a head-dress of black feathers.

They pressed round our vessel with much affability and confidence, but none of them ventured to accept our invitation to come on board. They had but little to give us in exchange for our goods, to which they pointed with eagerness, and received them with a kind of respect; only some cocoanuts, mostly unripe, some utensils which they had taken with them by chance, and their arms. These are long lances, made of cocoa-nut tree, at the bottom of which about a hand's-breadth of other wood is fastened with strings of cocoa-bast, and the point is either widened and two-edged, or single, and very long. They first hesitated to barter their arms, and would not part with them, except for long nails or scarlet worsted girdles. We obtained from them some fishing-hooks, which were two preces of real mother-of-pearl joined together, and wrought in the most tasteful manner, perfectly resembling those of the Sandwich islands.

Their boats are made of several pieces of wood well joined together with cocoa-bast cords. Both ends are rounded off, above the water and below the water, furnished with a projecting spar. They have an outrigger, on which their arms are secured.

We did not wait for a boat, which approached us under full sail, from a distant island of the group.

The low group of the Penrhyns richly provides for a numerous population, which is proved by the appearance of the people. Of their productions, we only know the incomparable cocoa-woods which cover them, and the pandanus. We have not been able to learn what other fruits they may have, or what roots; whether the swine and the dog, or whether the latter alone.

As we sailed from the Penrhyns, black clouds, with thunder and lightning, hung over them, and afforded us a sublime prospect, such as is seldom seen at sea.

THE LOW ISLANDS IN 15° S. LATITUDE, BETWEEN 138° AND 149° W. LONGITUDE.

ROMANZOFF ISLAND.

THE author of these papers leaves it to learned hydrographers, who require the most scientific research respecting the similarly formed reefs and low islands in these seas, on the one hand, to compare the low islands which we saw about 15° S. latitude, between 138° and 149° W. longitude from Greenwich, in the year 1816, namely, in the order in which they follow from E. to W. (the direction of our course,) the Doubtful Island (*Sumnitelny Ostroff,*) Romanzoff and Spiridoff islands, Rurick and Dean's Chain, and Krusenstern's Island, with the discoveries of former navigators, and especially with those of Le Maire and Schouten, whose course we followed; and, on the other hand, to look for their names on the chart of Tupaya, in the limits of which they are situated.

Krusenstern has treated on the first of these points in his Contributions to Hydrography, p. 173. But we cannot recognize in the desolate Spiridoff Island, that of Sondergrondt, which is said to be very populous, and richly covered with cocoa-trees; and this seems to cast a doubt on his decision upon other points.

The islands seen by us all appeared to be inhospitable, and really uninhabited, and the small Romanzoff Island was the only one of all those on which we landed that was covered with cocoa-trees. The formation to which they belong has already been described. We have only to add a few remarks on those we visited. A look at the atlas will be more instructive with respect to the others than all we could say about them.

Romanzoff Island is of small circumference. The raised dam of madrepore, which forms the outward edge, incloses a flat, where there seems to be a greater depth of mould, on which slender cocoa-trees rise here and there, without forming an entire wood. The raised protecting edge is partially broken on the leeward side of the island, and it appears, that in very high tides the sea penetrates into the interior of the island. The rainwater which was collected in many places was perfectly sweet.

The Flora here is extremely poor. We counted only nineteen species of perfect plants, (one fern, three *Monocotyledones* and fifteen *Dicotyledones*,) and we do not think that many have escaped our observation. The low *Acotyledones*, with which vegetation begins in higher latitudes, seems to be wanting here. The lichens appear only on the trunks of old trees, like a powdery covering, and the black mould on the stone does not seem to be of a vegetable nature. A moss, and several fungi which we found at Radack, did not occur in Romanzoff. The plants that we observed were, a *Polypodium*, the cocoa-tree, the pandanus, a grass, *Scaevola Konigii*, *Tournefortia argentea*, *Lythrum Pemphis*, *Guettarda speciosa*, a *Cassyta*, an *Euphorbia*, and a *Boerhavia*, a herbaceous nettle. Besides the above, all of which are at Radack, we met with the following, which are wanting there; two shrub-like *Rubiaceæ*, another shrub *Heliotropium prostratum*, *Portulacca oleracea*, a *Lepidium* (*acre?*), and a *Buchnera?*

Shrubs with entire, simple, and for the most part fleshy leaves, and colourless flowers, form an easily penetrable thicket, over which the cocoa-tree rises, and where the pandanus alone is distinguished by its remarkable form, and only the leafless *Cassyta* climbs with its reddish fibres. The soil is every where to be seen through the thin vegetable covering.

We did not observe the rat, which, it is true, conceals itself during the hot hours of noon, (at which time we visited the island.) Different kinds of forest birds (*Numenius*, *Scolopax*,) were numerous on this island, which did not appear to fear man. They merely avoided our treading on them like tame fowls in a farm yard. The *Sterna stolida* is the most numerous among the water-fowls. The foolish confidence of this bird has justly acquired its name. In these seas it literally flew into our hands, and we set some of them at liberty after we had tied a ticket round their necks, with the name of the ship and the date.

A small lizard seemed to be the only animal without wings on Romanzoff Island. A small butterfly was very common, and the only insect that fell into our hands.

Romanzoff Island is visited by the inhabitants of other islands, which are not to be seen from it. The landing-place is on the windward side. From thence several shining footpaths, trodden in the fragments of sharp coral, lead, in different directions, through the island. We found in the interior a small boat, which was quite decayed, made of the trunk of a cocoa-tree, hollowed out, and furnished with an out-rigger. In two different places stood light circular huts, consisting only of a few sticks, coarse mats, and cocoa-leaves. In one of them we found an instrument resembling a comb of wood, fastened together with strings of cocoa-bast. There were several pits dug for the collection of rain-water. Fire had been kindled on different parts of the ground, but we nowhere found baking-pits. Under the lee of the island, along the shore, was a place contrived to stretch lines; and near this place was a young tree, with the branches cut off, on which hung cocoa-nuts, and leaves, and a string of cocoa-bast.

There were no settled dwelling or morais on Romanzoff Island, and we no where found indications that people had lately visited it.

WAIHU, OR EASTER ISLAND.—SALA Y GOMEZ. *

We but barely set our feet on the lava beach of Easter Island, and do not flatter ourselves with considerably extending the knowledge the reader may already have of it. We refer to the reports of our predecessors, and only try to give him an account of the impression which the transient view of it made upon us.

Easter Island rises majestically out of the waves, in a triangular form, swelling into pyramidical mountains.

It represents in miniature the regular large lines of Owhyee. It appears to be covered every where with the liveliest green: the ground, even on the steepest declivities of the mountains, is divided in regular fields, which are distinguished by various lively colours, and many of which were covered with yellow blossoms. We gazed with astonishment at this volcanic earth covered with stone, which is notorious for its want of wood and water.

We believe we have distinguished on the south-east coast, by the aid of our telescopes, some of the

* Krusenstern's Contributions to Hydrography, p. 219.

colossel statues which have excited so much admiration. The busts in Cook's bay, on the west coast, where we cast anchor, which distinguished the landing-place, and were seen by Lisianskoy, no longer exist.

Two canoes (we only saw three in all), each containing two men, approached with signs of friendship, but without venturing near the ship. Numbers of the inhabitants swam round the boat, which was sent out to sound, offering to barter with it. A deceit of one of these traders was severely punished. To effect a landing, we had another boat put out. A numerous assemblage of peaceable people, friendly, noisy, impatient, and disorderly, awaited our arrival on shore. It is not our part to decide, with La Peyrouse, if these childish people are to be pitied, that they are more disorderly than their brethren. It is certain that this circumstance renders intercourse with them more difficult. We approached the shore. Every one ran about, rejoiced, and shouted; signs of peace, threatenings, stone-throwing, and shots, testimonies of friendship, were exchanged. Numbers of them at last ventured to swim round us in crowds, when trading began, and was conducted with honesty. All of them with the repeated cry of *Höe! Höe!* desired knives, or iron, for which they offered us in exchange fruits, roots, and very neat fishing-nets. We got on shore for a moment.

The people who had been represented so ugly,

appeared to us as having handsome features, an agreeable and expressive physiognomy, a well-made, slender, and healthy form, and old age not accompanied with infirmity. The eye of the artist rejoiced to gaze upon a more beautiful nature than the bathing places in Europe (his only school) afford him. The blueish broad-lined tattooing, which exactly follows the direction of the muscle, has a very pleasing effect on the brown ground of their skin. There seems to be no want of bast stuffs. White or yellow mantles of it are very general. Fresh wreaths of leaves are worn in their hair, which is cut of different lengths. Head-dresses of black feathers are more rare; we observed several ornamental necklaces which were adorned in the middle with a polished shell (*Patella*). We met with no inelegant, disfiguring ornaments. Some old men had their pierced and extended ear-lappets tied together, drawn again through the hole, and thus not very perceptible. The fore-teeth were frequently broken off. Some young men were distinguished by a much lighter colour of the skin. We saw only a few women, and these had their faces painted dark red, without either beauty or grace, and seemed to be without consideration among the men. One of them had an infant at the breast. We do not feel ourselves authorized on this account to draw a conclusion respecting the proportion of the numbers of the two sexes.

If we compare the reports of Cook, La Peyrouse,

Lisianskoy, with our own experience, a supposition arises that the population of Easter Island has increased, and that the situation of the islanders is improved. But if the generous intentions of the philanthrophic Louis XVI., who sent to this people, by La Peyrouse, our domestic animals, useful plants and fruit trees, have succeeded, we were not able to ascertain, and must therefore doubt it; we only saw the productions enumerated by Cook, as bananas, sugar-cane, roots, and very small fowls.

When we weighed anchor in the evening fertilizing clouds rested on the mountains of the island.

We have since learnt, the probable occasion of the doubtful reception we met with at Easter Island, and had cause to blush for ourselves who call these people savages.

The island Sala y Gomez is merely a naked, low rock, which rises like a saddle at both ends, where the kind of rock is exposed to view, the middle appearing to be strewed with congeries. It does not belong to the coral reefs, which only appear farther to the west. We may suppose a connection, and a similar nature with the high volcanic land of the neighbouring Easter Island. As yet there are no signs of the beginning of a future vegetation visible. It serves as the residence of innumerable sea-fowls, which seem to prefer these naked rocks to green though uninhabited islands, as the plants contain insects; and the ant is particularly dangerous to their young.

The sea-fowl are always met with. according to our frequent experience, to windward of the island, where they build their nests. In the morning they are seen flying far from the shore with the wind, and returning in the evening likewise with the wind. Kadu also seemed to notice the flight of the birds in the evening.

Fragments of a wrecked ship are said to have been seen at Sala y Gomez; we, however, looked for it in vain. One shudders at the idea of the possibility that a human being might have been cast here alive; for the eggs of the sea-fowl would have but too well sufficed to protract his wretched existence, between the sea and sky, on these bare sun-burnt rocks.

THE SANDWICH ISLANDS—THE JOHNSTONE ISLANDS.

OWHYEE rises majestically, in grand unbroken lines, from the waves, and forms, in an enormous mass, three different mountain-summits, on two of which the snow lies several months in the year.

We both times visited the Sandwich islands in the autumn, and never saw any snow on the heights of Owhyee. (In November, 1816, and in September, 1817.)

Mouna Roa, the great mountain, La Mesa, (the Table of the Spaniards *,) rises in a bold curve southwards, in the interior of the island, and towers above the others, which unite with it. Mouna Kaah, the little mountain, the next to the Mouna Roa, with its rugged cliffs, occupied the north. The third, Mouna Wororay, a volcanic peak, lies on the western coast. There is a drawing of the crater in Vancouver's atlas. On its naked declivities shine streams of lava, the last of which it

* Owhyee and the Sandwich islands, La Mesa, or La Mira, and Los Monges, of the old Spanish charts (San Francisco, of Anson's chart might perhaps likewise be Owhyee,) must have been frequently seen by the galleons on the voyage from Acapulco to Manilla. It is to be observed that M. Marin has not been able to discover, in the popular traditions of Owhyee, any reminiscence of a former intercourse with Europeans.

poured into the sea by a lateral eruption in the year 1801. The village of Powarua is built on the shore upon this scorious lava. The Mouna Puoray, which forms the north-west point of the island, joins, as an inferior hill, to the basis of Mouna Kaah.

The heights of Owhyee generally appear quite distinct during the night, and in the morning; the vapour falls on them at noon: the clouds which they produce repose in the evening in dense masses over the island, and dissolve towards midnight.

Where we approached Owhyee, doubling the north-west point, and sailing along the west coast to the southern foot of Woraray, near Titatua, the declivities appear bare and sun-burnt. Some parts are used for tillage, the most are covered with scanty grass. Amidst clouds, the region of the forests begins, and the eye scarcely reaches the naked crowns of this gigantic mountain. The strand presents to the view an uninterrupted row of settlements, which, as you approach further to the south, are surrounded with more luxuriant verdure, and more frequently relieved by cocoa-palms.

Of the volcanic chain of mountains of the Sandwich islands, only Wororay on Owhyee seems still active. Hot springs are in the territory of Kochala, near the habitation of Mr. Young, on the coast south of Puoray. The chain runs from the north-west point of Owhyee over the islands of Mowee, Morotoi, and Woahoo, to west north-west. The most eastern mountain on Mowee is but little inferior in height to Wororay, and is similar in the grandeur of its form. The more western is lower, and its summit appears to be rent in two different clefts from north to south.

The grand lines of the mountain descend on Morotoi still lower, even to the flat western point of this island. The mountain rises again in Woahoo, where, with an entirely different form, it scarcely reaches a quarter of the height of Owhyee. Two dissimilar groups of mountains rise in the island of Woahoo. The eastern and lower one has a greater extent than the western, which contains the higher summits. The chain, deeply indented by well-watered and beautifully verdant vallies, shoots up ragged summits in broken lines. The woods descend lower than in Owhyee, upon their declivities, to the sun-burnt plains, which mostly surround these islands, and were once coral reefs covered by the sea; and such reefs extend far into the sea from these plains. A break in the reef, at the mouth of a stream produced by water collected in the mountains, forms, on the southern foot of the eastern mass of mountains, the secure harbour of Hana-rura, from which place we made our excursions in different directions through both parts of the island.

The nearest low hill behind Hana-rura is an old

volcanic crater, now closed, and which, like the external declivity, is thickly overgrown with grass. A similar, but higher and larger crater, which forms a promontory, washed by the sea, bounds the prospect to the east. Pretended diamonds, which an European is said to have found in these parts, gave occasion to the taboo, which has been imposed upon this mountain. We were shown as such common quartz crystals.

The mountain-chain rises behind these naked front hills, covered with a lovely verdure in irregular steps to its highest ridge, which runs along the northern coast. Valleys and defiles lead to the passes which intersect each other between the summits. The valley of Nuanu, behind Hana-rura, is the most extensive and pleasant of all. Beyond, towards the north or north-east, the mountain presents a steep declivity, which cannot be ascended, except barefooted, by giddy paths and rocky ascents.

Low hills, covered with sun-burnt savannahs, unite the two mountain-masses of the islands. To the south of these hills, the inlet of the sea, called by the English the Pearl River, winds with numerous branches up to their foot, through an extensive plain, which is a coral reef abandoned by the sea, the surface of which is raised about ten feet above the present level of the ocean. This frith appears to afford the finest harbour; it is, however, said, that a bank obstructs the entrance of ships. It receives streams only from the eastern mountains. The ridge of the western higher mountains is turned to the interior of the island, and pours its waters into the valleys, which it embraces in the west, with several arms. The passes between the summits are high and steep, and only to be ascended by dangerous paths. The luxuriance of the vegetation, which, at the height of about three hundred toises, to which we ascended, appears unchanged, mostly conceals from the eye of the geologer the object of his search, and the kind of rock is scarcely to be seen.

On both sides of the island we observed only argillaceous porphyry and amygdaloid; black patches, which we noticed from the sea on the eastern declivity and foot of the great old crater, seemed to us to be a lava.

The clouds collect round the summits of the mountains, and rain frequently falls in the interior of the island, while a scorching sun burns up the strand.

The temperature visibly changes as soon as you enter the mountain-valleys from the exterior plains.

We already possessed three measurements of the height of Mouna Roa, differing very much from each other, after King, Marchand, and Horner. The more exact measurement of M. Von Kotzebue determines it within six toises of the second above-mentioned; and his trigonometrical labours on the other summits of the Sandwich islands, present an interesting series.

The short time which was allowed us on both visits, only permitted us to look with regret on the mountains of Owhyee, which appeared to us worth the object of a voyage on purpose to the Sandwich islands. We were even obliged to renounce the examination of them, though on the spot.

To ascend Mouna Roa from Titatua requires a journey of at least two weeks (compare Vancouver), and at Titatua and Powarua at the very foot of Wororay, the summit of which we could hope to ascend in a short time, there still remained the journey to the ship at Hana-rura, in a double canoe of the natives, which was not to be depended on, as you can have no command at all over such a boat; the navigation is impeded by frequent taboos, and the passage from Owhyee over to Mowee, and from Morotoi to Woahoo might be long delayed and prevented by the wind. What Archibald Menzies, the learned companion of Vancouver, has colleeted in plants on different excursions to the heights of Owhyee and Mowee, lies still buried with so many other treasures in the Herbarium of Banks; and though the venerable Nestor of natural philosophy has opened his Gazophylacium to the naturalist, and to all learned men, with the same unequalled hospitality, no one has yet undertaken to make us acquainted with the Alpine Flora of Owhyee.

The Flora of Owhyee has nothing in common with the nearest continent of the coast of California. The leafless form of the *Acacia*, the species of *Metrosideros*, *Pandanus*, *Santalum*, *Aleurites*, *Dracaena*, *Amomum*, *Curcuma*, *Tacca*, impress the stamp of its origin and natural relationship. Predominant are the families of *Rubiaceæ*, *Contortæ*, and *Urticæ*, of the latter of which several species, growing wild, are used to prepare different kinds of bast-stuff.* Several arborescent, milky *Lobeliaceæ* are distinguished. The exterior boundary of the island produces but very few kinds of grass and shrubs. In the interior the Flora is rich, without, however, being comparable with the luxuriant abundance of the Brazilian soil. Only low trees descend to the valley; among them the *Aleurites triloba*, with its whitish foliage, forms conspicuous thickets round the skirts and on the declivities of the mountains. You find here and there, in the high mountain defiles, wonderful banana groves, the trunks of which, crowded close together, cause a gloomy night, with their broad, extended foliage. This plant, which, cultivated on the strand, reaches scarcely a height of five feet, attains in these places thrice that height. The acacia, out of the

* The paper-mulberry tree, (*Broussonetia papyrifera*,) is cultivated in the Sandwich islands, as in most of the islands of the South Sea for making stuffs. But it is an erroneous supposition that stuff is prepared only from the bark of this tree.

trunk of which the large canoes of the natives are hollowed, attains only in the mountains the height necessary for this purpose; you find also here and there the sanders-tree, the wood of which, so highly valued in China, bestows riches on the rulers of these islands, while the oppressed people, who are obliged to collect it, are impoverished, as they are taken away from their agriculture and arts.

The taro-root, (*Arum esculentum*), beaten to a soft pap, after it has been boiled, constitutes the chief food of the people. Woahoo is the most fertile of the Sandwich islands, from which Owhyee receives a part of the taro necessary for its consumption. The cultivation of the valleys behind Hanarura is remarkable. Artificial ponds support, even on the mountains, the taro plantations, which are at the same time fish-ponds; and all kinds of useful plants are cultivated on the intervening dams. Many imported plants are now cultivated close to the originally native, but the people who are attached to their ancient mode of life, make use of only a few of them. Among these the tobacco must be chiefly mentioned, the use of which all the nations of the earth have been equally ready to adopt. The water-melon, the melon, and fruit in general, have, next to tobacco, met with the best reception. Besides the pernicious kava, fermented liquors are prepared from the tea-root, (*Dracaena terminalis;*) but the sugar-cane is not yet employed for this purpose.

To the diligence of M. Marin, as a farmer, the Sandwich islands in general, and Woahoo, his present residence, in particular, lie under great obligations. He has assiduously introduced and multiplied our various kinds of animals and plants. He has near Hana-rura numerous herds of cattle. (Goats seem to be more generally spread). He possesses horses, and will increase the breed of asses and mules, which are more useful in these mountains. Many foreign trees and plants are reared in his plantations. Several which he has introduced, are already found everywhere growing wild; for example, the *Portulacca oleracea,* (only two other species of the same kind belong to the native Flora). He has lately had rice grow from Chinese seeds, after many fruitless attempts. He has planted vineyards of considerable extent, and the grapes flourish very well, but he is yet unskilled in the art of making wine. We have frequently had occasion to remark, during the course of our voyages, that the art of using the productions already existing, is a more urgent want than the introduction of new ones, and embrace this opportunity of giving a useful hint to beneficent travellers. There are only wanting a few books of instructions.

The only original wild quadrupeds of the Sandwich islands are a small bat and the rat. To these is added our common mouse; besides the flea, some species of blatta, and other noxious parasites.

The oxen are now grown wild in the interior of Owhyee, where the king sometimes has them killed for his table. We observed among the land-birds the *Nectarina coccinea*, whose highly-valued feathers form a part of the tribute. The sea is rich in fish, many kinds of which are adorned with colours of extraordinary splendour. They are numbered among the favourite dishes of the inhabitants who breed them in the taro plantations, and in fish-ponds which are formed by places walled in on the reefs along the strand.

Among the crabs, the beautiful *Cancer squilla* and *Palinurus* species are distinguished; among shells the little pearl oyster, which are only caught in the Pearl River, and from which small pearls, of little value, are obtained.

The sea-worms and zoophytes, probably compose the richest and most interesting part of the Fauna. In general, the species here appear to be different from those at Radack. The progressive growth of the reef does not seem to have escaped the natives. They told us that the men, who, at the king's order, fetched stones out of the sea, to build a wall, declared, while at their work, that it would grow, and increase of itself.

Respecting the Sandwich islands, we possess only the accounts of superficial travellers, who, in their fidelity, place images before us, where we expect, and are always more excited to desire sounder knowledge. Cook discovered these islands;

and an unhappy quarrel, caused his valuable life to be terminated among the powerful and warlike Owhyeeans. They adored him as a god, and still piously revere his memory. Trade followed the traces of Cook to the north-west coast of America; and the Sandwich islands, which afforded all kinds of refreshments to the navigators sailing thither, thus received the importance which their discoverer ascribed to them. With Vancouver we are at home with them. A great man, with whom, while yet a youth, we were made acquainted by Cook, seized the reins of power at Owhyee, and strove for the sole government of the whole group. Tamaahmaah assured himself of the protection of Great Britain, by freely and voluntarily doing homage to King George, in the hands of his friend Vancouver. Later travellers, down to Lisanskoy, who were informed by the Europeans settled on the Sandwich islands, extend our knowledge of them, and acquaint us with the sequel of the history. Our rapacious adventurers diligently promote war, to keep up the price of arms, which they give in payment. Tamaahmaah accomplished the conquest of all the islands, and the king of Atooi, (the group lying apart in the west), hastened to submit voluntarily to him whom he could not resist. He was, it is true, misled into rebellion, under the standard of the Russian American Company, but he immediately atoned for his fault, and did homage anew to his sovereign, (1817).

Tamaahmaah, favoured by the situation of his kingdom, and the sanders-wood, which it produces, has collected immense riches. He buys, with ready money, arms and ships; builds himself smaller vessels, which when he spares the copper-sheathing, are drawn on shore, and preserved under sheds at Titatua, Karakakooa, and other places on the island of Owhyee. He sends out his ships manned half by Europeans, and half by natives; and tries (in which he has not yet succeeded), to procure admission for his flag into Canton. He chooses with great judgment such Europeans as offer him their services, and to those whom he employs, he is very liberal in lands and salary; he is noble-minded, and, with the instruction he receives from foreigners, remains still faithful to the spirit of his people, and to the customs of his ancestors.

But after the death of the old hero, his kingdom, founded and kept together by force, will fall to pieces, the partition of it being already decided upon, and prepared.

Kareimoku, otherwise called Naja (Bill Pitt of the English), descended from the royal blood of Mowee, being still a child on the conquest of this island, was spared, kindly treated, and educated by Tamaahmaah. He has given him his affection, lands and power; raised him to an elevation, scarcely inferior to his own. He has given him the power to decide on life and death; and has always found him faithful. Kareimoku, governor of Woahoo, and lord of the fortress of Hana-rura, in this island, which is the most important of all, on account of its harbour, is prepared to take it for himself, and buys ammunition and ships on his own account. He is perfectly agreed, and bound by intimate friendship, with Teimotu, of the war race of Owhyee, and brother of queen Kahumanna, who is to have the island of Mowee for his share. The king of Atooi will retain his hereditary kingdom, as an independent prince; and the natural heir to the kingdom, the weak, dull-minded Lio-Lio, (the Prince of Wales of the English,) nephew of the last king of Owhyee, son of Tamaahmaah and the lofty queen Kahumanna, before whom his father only is allowed to appear uncovered, will be limited to the hereditary island of Owhyee. No foreigner, though there are so many among the most powerful chiefs, and vassals of the kingdom, can make any claim to govern the natives.

Notwithstanding these approaching political changes, the Sandwich islands will remain what they are — the free port and staple of all the navigators of these seas. But should any foreign power conceive the foolish idea of taking possession of them, the jealous vigilance of the Americans, who possess the almost exclusive commerce of these seas, and the secure protection of England, would not be wanting to frustrate the undertaking. The conquest might indeed be effected. The fort in the back-ground of the harbour of Hana-rura, which Mr. Young has erected without judgment, is merely a square of dry brick wall, without bastions or towers, and without ditches; and does not answer the double intention of the Governor, to defend himself against an external attack, and an internal enemy. The fort ought to be regularly built where it now stands, and there ought to be a battery on the external edge of the reef, to defend the entrance of the harbour. Notwithstanding their stock of ammunition and arms, the natives are not yet acquainted with the management of artillery, or with our military art. A serious invasion might appear to be decisive; but the conquerors would only have conquered the earth for their grave. This people will not submit to strangers; and it is too powerful, numerous, and martial, to be quickly extirpated, like the natives of the Mariana islands.

This is the political situation of the Sandwich islands. What is affirmed in the Missionary Register of 1818, p. 52, that a son of Tamori, king of Atooi, who is at present educated in the school of the foreign missions in Cornwall, (Connecticut, North America,) with other Owhyeeans, is the real heir of all the Sandwich islands, betrays inconceivable ignorance.

No missionaries had yet come to the Sandwich islands; and, in truth, they could promise themselves but little fruit among this sensual people. Christianity cannot be established in Eastern Polynesia, but on the overthrow of every thing existing. We do not doubt the events at Otaheite, but, at the same time, we cannot conceive them; and M. Marin, who had previously visited these islands, told us, what is very clear, that the natives, for the most part, only visited the missionaries to have the pleasure of mimicking their customs.

We are indebted to the contributions of William Mariner, and to the laudable diligence of Dr. John Martin, for the most valuable addition to the knowledge of Polynesia, in the satisfactory "Account of the Natives of the Tonga Islands, London," 1818. This important work was not published at the time of our voyage; and therefore the want of a similar work on the Owhyeeans; the desire thoroughly to study the traditions and the history, the common and religious language; the religion and customs, the social habits, and spirit of this people, as well as the eager wish to examine, on the mountains of Owhyee, the history of the plants, and their migrations, urgently induced the naturalist of the expedition, on our first visit to the Sandwich islands, to offer to remain there till the return of the Rurick. This idea, which would in the end have been baffled, by the present political circumstances, was deemed incompatible with the object of the expedition. It is now time, under the protection of the noble-minded Tamaahmaah, and with the assistance of the Europeans settled in his kingdom, whose experience and knowledge would be of great advantage to the learned inquirer, to undertake this work, to commit to writing what the Owhyeeans know of themselves; for where monuments and letters are wanting, languages change under foreign influence, traditions are lost, manners assimilate, and the European will one day find on the Sandwich islands, new Europeans, who will have forgotten their origin and their ancestors.

Of all the Europeans settled there, M. Marin appears to possess the most extensive knowledge of the people of Owhyee. He has studied it in many respects and had occasion to compare and enrich his observations on many voyages to other islands in the South Sea, from Otaheite to the Pelew islands. M. Marin had committed them to paper, and we regret, with him, the loss of his MSS. During our first stay at Hana-rura, he promised to answer in writing several questions we had put to him, and to give us his remarks on our return. But we were deceived in the hopes he had given us reason to entertain. He had not been able to spare time for this work, and during our second stay, he was so much engaged with the ships lying in the harbour, that we were able to enjoy his instructive conversation only for a few moments.

M. Marin regretted the recent death of an old man at Woahoo who was particularly acquainted with the ancient traditions of his people, and with whom a part of their traditionary history is probably lost. These ancient traditions are very differently related. There was a flood which covered every thing, and only the summit of Mouna Roa rose above the waves. The people saved themselves upon it. Previous to this flood there had been another revolution, by which the earth was darkened for forty days. Formerly strangers, whose names are mentioned, arrived in a boat at the Sandwich islands. M. Marin had heard a tradition at Otaheite, according to which mariners of this island, who had been lost at sea, were the very same who had been thrown upon the Sandwich islands.

The relations of social order which are not founded on written rights and laws, but upon faith and custom, which are more powerful than force, may be considered and interpreted in different ways. M. Marin assumes four casts in Owhyee. The princes, the nobility, the middling class (who constitute the great majority of the people) and the populace, a despised class, which is not numerous. Formerly every white was considered as equal to the nobility, now his rank depends upon his character.

The word *Hieri, jeri, erih, ariki,* or *hàriki* (chief), is best to be translated by lord. The king is *Hieri ei Moku,* the Lord of the Island or Islands. Every powerful prince or chief is *Hieri Nue,* Great Lord, and by this title Tamaahmaah,

Kareimoku, Haulhanne (Mr. Young), are called without distinction.

The land belongs to the lord of the island, the nobles possess the ground only as fiefs; the fiefs are hereditary but inalienable, they return to the king. Powerful nobles may perhaps rebel and defend what they possess. The right of the strongest makes the lord of the island. The great nobles decide their private feuds by arms. These petty wars, which were formerly very frequent, seem to have ceased since the year 1798. The noble leads his people to war: no one who is not a noble can possess a fief or lead people in war. He can only be the steward of the estate. Those who cultivate the land are farmers or peasants of the feudal lord, or immediately of the king. The king receives tribute from all the land. Distinguished chiefs are placed as governors over different islands and territories. The people are almost subject to the arbitrary will of the lord, but there are no slaves or vassals (*Glebæ adscripti*).

The peasant and the labourer may go wherever they please. The man is free, he may be killed, but not sold and not detained. Lords and nobles without land serve those who are more powerful. The lord of the island keeps many of them, and his rowers are exclusively of this class. It is to be understood, that the casts are so determined, that it is not possible to go from one to the other. Nobility which can be given and taken away is none. The woman does not share the rank of the husband. The rank of the children is determined by very fixed and certain laws, chiefly after that of the mother, but also after that of the father. A noblewoman who marries a man of the lower class loses her rank as soon as she has children, in which case she and her children descend to the cast of the husband. Primogeniture does not determine hereditary right, but in the plurality of wives the more noble birth by the mother's side. The inequality of the nobility, and the different degrees of the taboo or sanctity which is due to every distinguished chief after his birth, without respect to his power, were not sufficiently explained to us. The predecessor of Tamaahmaah, in Owhyee, was taboo to such a degree that he was not allowed to be seen by day. He only showed himself in the night: if any person had but accidentally seen him by daylight he was immediately put to death; a sacred law, the fulfilment of which nothing could prevent. The human victims, who are here killed at the death of the king, princes, and distinguished chiefs, and buried with their remains, are of the lowest class. In certain families of this cast the fate of dying with the different members of such or such a noble family is hereditary, so that it is known at the birth of a child, at whose death he is to be sacrificed. The victims know their destination, and their lot does not seem to have any terror for them. The progressive spirit of the times has almost made this custom antiquated, which will hardly be repeated at the death of the most sacred person. When three victims, on the death of Kahumanna's mother, offered themselves to fulfil their destination, Kareimoku would not allow it, and no human blood was shed. Human sacrifices, it is true, still take place, but it would be unjust to upbraid the Owhyeeans for them. They sacrifice culprits to their gods, as we sacrifice them in Europe to justice. Every land has its peculiar customs. What were the Christians when *autos-da-fe* were celebrated, and how long have they ceased? The custom of eating human flesh had ceased long before the death of Cook. The last historical traces of it may be found in the island of Woahoo.

Every great chief has his peculiar gods (*Akua*), the idols of which are represented in his morai. Others have different ones. The worship of these idols appears to be more for distinguished parade than religion. The common people must do without these idols, and they make various creatures, birds, fowls, &c. for the object of their worship. Superstition prevails under many forms in the Sandwich islands. As Kareimoku's guests, we were present at the celebration of a *Tabu pori,* which lasted from the setting of the sun to sunrise on the third day. It is already known what degree of sanctity is imparted to him who joins in this communion with the gods during the time. Should he accidentally touch a woman, she must be instantly put to death. Should he enter a woman's house, the flames must immediately consume it. We expected a certain seriousness during these sacrifices and prayers; and were astonished at the profane disposition which manifested itself; the indecorous sport that was made with the idols; and the tricks which they delighted to play us during the sacred ceremony. Children show more sedateness in playing with their dolls.

The restrictive laws of the taboo*, otherwise remain in undiminished power. We ourselves saw the corpse of a woman floating round our ship, who had been killed, because she had entered the eating-house of her husband in a state of drunkenness. It is, however, said that the women, when they know they are unobserved, make no difficulty of transgressing the many prohibitions to which they are subject. The intercourse with the Europeans has not yet had any particular influence on

* They are known by the voyages of Cook, Vancouver, Turnbull, Lisianskoy, &c. Three houses necessarily belong to one family; the eating-house of the men is prohibited (taboo) to the women. The residence is common: the women's house is not closed against our sex, but a decorous man will not enter it. Each sex must dress their own victuals, and over a peculiar fire. On ship-board the prohibition (taboo) is less rigorous. The two sexes are not allowed to partake of the flesh of the same animal. Hog's-flesh, (not that of dogs, which is equally valued,) turtle, as also several kinds of fruit, cocoa, banana, &c. are taboo for the women. The male attendants of the women are in many respects subject to the same restrictions as themselves.

the social order, arts and manners of these people. We have certainly contributed to strengthen in them only the vices and arts of corruption which are revolting in these simple people. *Ingens nostratium Lupanar! Turpissimis meretricum artibus, fœtidissimis scortorum spurcitiis omnis instructa est fœmina vel matrona. Omnis abest pudor, aperte avideque obtruditur stuprum, precio flagitato. Aperte quisque maritus uxorem offert, obtrudit solventi.*

A circumstance which occurred about the year 1807, is differently related by report. We follow the account of M. Marin.

A nephew of the king had been found in the arms of the queen Kahumanna. He himself escaped, but his garment, which he left behind, discovered him. Three days after this deed, he was seized and strangled by the nobles of the kingdom. A soldier on duty announced to the king the punishment and crime at the same time. This was the regular order. Tamaahmaah regretted the poor youth, and shed tears.

Compared with our friends in Radack, we found the Owhyeeans selfish, inelegant, and uncleanly. In their intercourse with strangers, from whom they hoped to derive advantage, they have lost their natural hospitality. Their great talent is mimicry, and habit made it very easy for us to understand each other. They are an incomparably stronger people than the Radackers. Hence arises greater self-confidence, and more unreserved cheerfulness. The chiefs are, in particular, of the handsomest and most robust make. The women are handsome, but not attractive.

Former travellers have observed, that, on the Sandwich islands, natural deformities are more frequent than on the other islands of Eastern Polynesia. We saw at Woahoo several individuals with crooked backs, an idiot, and several people of one family with six fingers on the hands.

The Owhyeeans are tattooed very little, and irregularly. It is remarkable that this national ornament has borrowed foreign patterns. Goats, muskets, even letters of the alphabet, name and birth-place, are frequently tattooed along the arm. The men shave their beards, and cut their hair in the form of a helmet, the crest of which is often bleached, so as to be of a light or whitish colour. The women have it cut short, keeping only a rim over the forehead, bleached white with unslacked lime, and standing up like bristles. Sometimes a fine long curl is preserved on the middle of the forehead, and dyed violet, which is combed back. To please the Europeans, some suffer their hair to grow, and tie it up behind in a *queue*, resembling that which was ordered in the regulations of the Prussian army, in 1800. The Owhyeeans have in general wisely remained faithful, both to their national costume, and to their mode of life. Their princes only dressed themselves in honour of us, with the utmost neatness, in fine English clothes, and imitated our manners with much propriety. They are at other times always dressed in their native costume, and only their foreign guests are served in porcelain and silver. Fashion reigns even at Owhyee, with varying caprice, particularly over the women. The ornaments which the queens and great people wear, immediately increase extremely in value. Every body has now a looking-glass, and a pipe-head tied round the neck in an European handkerchief. The Europeans are dressed in the European fashion, and do not uncover before those whose rank otherwise requires this mark of respect.

Many Owhyeeans understand a little English, but none are proficients in it, not even those who have made voyages on board American ships, which a great many have done. None of them have probably learnt the letters.* It is only our ships which attract their whole attention. We were very much surprised to see, at Titatua, some children drawing ships with a switch in the sand on the beach. Two and three masted vessels were drawn with the greatest accuracy, and provided with the

* Tamaamaah understands English without speaking it. Lio-Lio learned to write two lines in English, in which he begged the captain of a ship to send him a bottle of rum. Louis XIV. when a child, learned to write: "L'homage est dû aux Rois, ils font ce qu'il leur plait."—(MS. of Dubrowski, collection in the Imperial Library at St. Petersburg.)

most minute parts of the tackling. The Owhyeeans, however, build their boats according to the ancient mode, single or double. Larger double canoes, belonging to the king, which serve for communication between the different islands, are rigged after the European fashion. We must not confound, with Zimmermann, (in his Australia,) the boats of Eastern Polynesia (the Friendly, the Sandwich islands, &c.) which go with oars, and with sails only before the wind, with the ingenious vessels of the islanders of the first province (the Ladrones, &c.) which go with sails only, and with all winds. We have been made sufficiently acquainted with the former by Cook and other modern navigators, and with the latter by Dampier, Anson, &c.

Besides navigation, the warlike Owhyeeans take a pleasure in their weapons, especially their lances. They delight in warlike games, which are not free from danger, and even, when children, practise throwing the dart. The favourite play of the boys and youths, is to contend in throwing short and light reeds with which the wind plays, with certainty at an unsteady mark, and seems to allude to this weapon. They have but few other sports. Their game at draughts, which was found among them, has been superseded by the European draughts.

Poetry, music, and dancing, which, in the South Sea islands, appear hand in hand, in their orignal union, to adorn human life, deserve to be particularly attended to. The spectacle of the Hurra, the festive dances of the Owhyeeans filled us with admiration.

The words mostly celebrate, like the Pindaric Odes, the fame of some prince. Our knowledge of the language was not sufficient to judge of their poetry. The song is in itself monotonous. With the accompanying beats of the drum, it measures the turns of the dance, bearing, as it were, upon its waves a superior harmony. In the varying dance, the human form develops itself to this measure, in the most admirable manner, representing itself in a constant flow of easy unconstrained motion, in every natural and graceful position. We fancy that we see the antique starting into life; the feet only bear the dancer. He moves forward with composure. His body, his arms, all his muscles, are expressive; his countenance is animated. We fix our eyes upon him as upon the Mime when his art transports us. The drummers sit in the back ground, the dancers stand before them in one or more rows; all join their voices in the chorus. The song is at first slow and piano, and is gradually and regularly quickened and strengthened, as the dancers advance, and their action becomes animated. All execute the same motions. It is as if the same dancer stood several times repeated before us. These festal games of Owhyee remind us of the chorus of the Greeks, of tragedy before the dialogue was introduced; and, if we cast a look upon ourselves, we perceive into what a wrong path we have absurdly strayed, by reducing the dance to a motion of mere pleasure. These games intoxicate the Owhyeeans with joy. Their usual songs are danced in the same spirit, standing or sitting; they are of very different characters, but always accompanied by graceful motions of the body and the arms. What a school is here opened to the artist! What an enjoyment is here offered to the amateur!

This fine art, the only one of these islanders, is the flower of their life, which is consecrated to enjoyment and to pleasure. They live for the present moment without calculation of time, and an old woman knows no more of her age than that she has lived beyond the first period of enjoyment, beyond the age of twelve years.

The Owhyeeans are generally included in the accusation which our navigators make against the islanders of the South Sea in general, that of being addicted to stealing. That we have no reason to join in this complaint is probably to be ascribed to the protecting influence of Tamaahmaah, whose disinterested and noble mind honoured us as the successors of Vancouver. The Europeans settled here give honourable testimony to the honesty of the natives. They leave their doors and shops unlocked without apprehension. These people only commit theft on rich strangers, on board well loaded ships. How can we expect that our abundance of iron, this precious metal, should not excite the cupidity of the islanders of the South Sea? "*Why beholdest thou the mote that is in thy brother's eye, and perceivest not the beam that is in thine own eye?*" We do not here allude to the early ages of the conquests of the Spaniards, but we have before our eyes what deeds rapacious adventurers have committed even in our days, in these seas, where our laws cannot reach them. We have touched upon many of them in our pages; others are enveloped in the gloomy veil of night. It is our duty to be the advocates of the weaker party. Let our testimony be rejected, but let the accounts of all mariners, who have navigated these seas since they were opened to our trade, be impartially examined, from Vancouver's voyage to Nicolas's New Zealand. The reader will judge for himself. While we condemn and punish, men of our own colour, unjudged and unpunished, exercise kidnapping, robbery, cunning, violence, treachery, and murder. Sciences and arts have given us this power over our weaker brethren.

The commerce of these seas is said to employ two hundred North American ships, which number appears to us, however, exaggerated. The principal branches of it are the smuggling trade on the Spanish coast of both Americas, which is carried on, upon the Spanish side, by the monks, the fur-trade of the N. W. coast, the exportation of the furs collected in the Russian American factories, the sanders-wood of the Sandwich, Fidji, and other islands. The field is opened for the boldest adventurers. They attempt, they pursue new discoveries; (we remind our readers of the ship, which, according to Mackenzie's account, was seen about the year 1780 in the Icy Sea;) they take Aleutians or Kadiackers to catch the sea-otters on the coast of California, &c. Canton is the general market, Hana-rura a free port and staple place. The captain for the most part manages the business, and those disputes are not to be feared, which frequently happen between the captain and supercargo, when these two offices are separated. In the dangerous trade on the N. W. coast there is no good faith on either side, and they must be on their guard against the arms which they sell. Neighbouring tribes are very often engaged in war. They agree with the leader of one, and deliver to him his enemy, whom they seek to seize by cunning or violence, for a fixed price. They entice chiefs on board, kidnap them, and restore them to liberty for a certain ransom, &c. Men also whom they purchase on the southern coast, they sell to advantage on the northern. We have mentioned the kidnapping in the South Sea islands in our article on Guahon. It was an American, who, in an island along the coast of California, had all the male inhabitants driven together, and shot. Captain Door, (with the Jenny from Boston,) touched at Guahon in 1808, after having taken in a cargo of sanders-wood in the Fidji islands. He praised to Don Luis de Torres the kind and hospitable reception he had met with from the natives. In 1812 he made the same voyage in another ship. On his return, he told Don Luis how hostilely he had been received this time, and that he had lost a mate and four sailors. The natives told him, that in the course of time they had become acquainted with the whites, and had resolved to show no mercy to them. (Respecting the Fidji islands, see Mariner's Tonga islands.)

In the burying-place of the Europeans, near Hana-rura, we read this simple monument on Mr. Davis.

The remains
of
M. Isaac Davis,
who died at this
Island, April 1810,
aged 52 years.

When we last sailed from Hana-rura, we left Mr. Young sinking under the infirmities of old age. Both friends, whose united names have for a long time been distinguished in the history of these islands, will repose together. The children of Mr. Young, though heirs to his estates, will be lost in obscurity among the people, as they were not born of a noble mother.

The islands which Mr. Johnstone discovered in 1807, on board the frigate Cornwallis, in the W.S.W. of the Sandwich islands, and which we looked for in the spring of 1817, are, like the island of Sala y Gomez, perfectly naked rocks, which do not seem to belong to the formation of low islands. The reefs which are united to them form shoals extending to a great distance from them, which are very dangerous to vessels that navigate those seas.

METHODS OF KINDLING FIRE.

There are various methods of producing fire.

In the Caroline islands, a piece of wood being held fast on the ground, another short piece, about a foot and a half long, of the thickness of a thumb, even as if turned, and with the end bluntly rounded off, is held perpendicularly over it, and put in motion between the palms of the hand, like the mill used for making chocolate. The motion is at first slow, but is accelerated, and the pressure increased, when the dust produced by the friction collects round the borer, and begins to be ignited. This dust is the tinder, which takes fire. The women of Eap are said to be uncommonly clever at this process.

In Radack and the Sandwich islands they hold on the under piece of wood, another piece a span long, with a blunt point, at an angle of about thirty degrees, the point of the angle being turned from the person employed. They hold the piece of wocd with both hands, the thumbs below, the fingers above, so that it may press firmly and equably, and thus move it backwards and forwards in a straight line, about two or three inches long. When the dust that collects in the groove, produced by the point of the stick, begins to be heated, the pressure and the rapidity of the motion are increased.

It is to be observed, that in both methods, two pieces of the same kind of wood are used; for which purpose some of equally fine grain, not too hard, and not too soft, are the best. Both methods require practice, dexterity, and patience.

The process of the Aleutians, is the first of these methods, improved by mechanism. They manage the upright stick in the same manner as the gimlet or borer, which they employ in their work. They hold and draw the string, which is twice wound round it, with both hands, the upper end turning in a piece of wood, which they hold with their mouth. In this way, I have seen a piece of fir, turned on another piece of fir, produce fire in a few seconds; whereas, in general, a much longer time is required.

The Aleutians also make fire, by taking two stones, with sulphur rubbed on them, which they strike together over dry moss, strewed with sulphur.

KAMTSCHATKA,

THE ALEUTIAN ISLANDS, AND BEERING'S STRAITS.

We have cast a look over the waters of the Great Ocean, and its shores, and viewed the islands situated in it, between the tropics, from the Indies, as from the mother country to which they belong, and whence organic nature and man have spread over them.

We now turn from those gardens of pleasure to the dreary north, in the same ocean basin. The song is past. A clouded sky receives us on the very limits of the northern monsoon. We penetrate through the gloomy veil, which eternally hovers over these seas, and shores not shaded by a tree, inhospitably frown upon us, with their snow-crowned summits.

We shudder to find man also settled here!* That part of the land and sea, which we are now

* Homo sapiens habitat intra tropicos palmis lotophagus, hospitatur extra tropicos sub novercante Cesere carnivorus.—*Lin. Syst. Nat.*

Ipsos Germanos indigenas crediderim.—Quis—Asia aut Africa aut Italia relicta, Germaniam peteret? Informem terris, asperam cœlo, tristem cultu aspectuque, nisi si patria sit.—*Tacit. Germ.* 2.

about to describe, embraces the chain of the Forelands, which bound the waters of the Great Ocean to the north; and the seas, islands, and shores, which are to the north of it.

This chain extends from the peninsula of Kamtschatka, on the Asiatic side, by the Aleutian islands, to the peninsula of Alaska, on the American side, over which peninsula the volcanic mountainous shores reach to the continent of the New World. We include among the Aleutian islands, the whole island-chain, without entering into their divisions; and we also include the small volcanic islands of St. George and St. Paul lying out of the line, near to the north of Oonalashka, which are by some unaccountable means left out on Arrowsmith's charts, though they are perfectly known even to English navigators; for example, Sauer. In the north of the Forelands we met only with primitive mountains, ice, and alluvial sand. *

The coasts of the two continents run opposite each other; the Asiatic in a north-eastern, the American in a northern direction, and form between high promontories, the Asiatic East Cape, (Vostotschin-oi, or Tschukutskoi-noss,) and the American Cape Prince of Wales, the strait called

* On the northern coast, which we visited, we observed no trace of the floetz formation, which is entirely missed in the highest northern latitudes of Europe. The expedition of Captain Ross has placed the existence of floetz-lime, in Baffin's Bay, beyond doubt.

Beering's Straits. The basin, which is inclosed in these coasts and the Aleutian islands, is called the Kamtschatkan Sea. The island of St. Matwey, (Choris Island,) lies in the middle of it.

The Asiatic coast is high, and washed by a deep sea. It is indented on the north by the broad and deep gulf of Anadyr, which is bounded on the north side by the projecting Tschukutskoi-noss (Anadirskoy-noss). It is also indented between the noss (or point,) and Cape East, by Matschickma, and St. Lawrence bays. Near to the Tschukutskoi-noss, and to the south of the strait, lies the island of St. Lawrence, (Clerke's Island,) placed before the promontories, which are the pillars of the entrance, like a half moon before two bastions. The sea has more depth between the island and the Tschukutskoi-noss, than between it and the American coast, on which side the passage is broader and shallower. The eastern part of the island appears to be a group of rocky islands, which the low alluvial banks have united into one. Some inaccessible rocky islands rise out of the sea, between St. Lawrence Island and Beering's Strait, and in the middle of the strait itself.

The American coast is inaccessible between Bristol Bay on the south, (to the north of the peninsula of Alashka,) and Norton Sound on the north, which, by its situation, corresponds with the Gulf of Anadyr, on the opposite Asiatic coast. The sea has no depth, and the waves break before you are in sight of land. A considerable stream from the interior of America is said to empty itself here, and to form sand-banks on the coast.

We penetrated through Beering's Straits to the north. The two coasts retire from each other. Cook saw the Asiatic coast, as far as the North Cape, under 68° 56′ north latitude; the American, as far as the Icy Cape, 70° 29′ north latitude. Low alluvial ground forms the shore before the high lands of America, and the sea which washes it has no depth. According to Cook, the Asiatic coast is of the same nature. The land seems to gain upon the sea by the alluvion of sand; and it is to be feared that it will gradually fill up this sea.

The sandy coast of America is indented by many inlets and friths. We left the southern Schischmareff Bay unexplored, and penetrated into the broad Kotzebue's Sound, which, to the south of the high Cape Mulgrave, runs in a south-eastern direction into the primitive land, and the back of which approaches to Norton Sound, which runs in from the southern part of Beering's Straits. * A frith, which opens in alluvial land, on the southern side of Kotzebue's Sound, and leads into the open sea in a voyage of nine days, in the baydares of the natives, which we call the Bay of

* Compare the accounts collected by Kobelef, in 1779, among the Tschukutskoi, and the later Russian charts, which Arrowsmith and other geographers follow.

Good Hope, may probably unite both, and divide Cape Prince of Wales from the continent, and make it into an island; for this inlet seems too near to Schischmareff Bay, to suppose that its course, as described by the natives, can lead into the latter.

To the north of Beering's Strait, lies before us the still unexplored field of the last important problems in geography; and we are called upon to give our opinion on them at a time when several expeditions are fitting out to examine into the facts, and our voice expires unheard. We proceed with hesitation to this task.

Are Asia and America separated? and is the sea into which you penetrate through Beering's Strait to the north, the great Icy Sea itself? or, is this basin a Bay of the Southern Ocean, bounded and surrounded by the coasts of the two uniting quarters of the globe in the north?

Can a north-west passage be possible, from the waters of Hudson's and Baffin's Bay, along the north coast of America to Beering's Strait?

Can it be possible to come into Beering's Strait from the Atlantic Ocean, northwards of Spitzbergen, and even over the north pole itself? and is there an open navigable polar sea, or a polar glacier of firm solid ice?

A man, whose name inspires us with the greatest respect, who is equally distinguished by learning and sound criticism, who was himself a companion of Cook in his second and third voyage, who has repeatedly navigated the South Polar Ocean, and the sea to the north of Beering's Strait, James Burney, is inclined to suppose that Asia and America are united, and parts of one and the same continent. *

We confess that Captain Burney has not gained us over to his opinion. In his Chronological History of North-Eastern Voyages, we find the historical testimonies that bear upon this question treated with the greatest freedom, and refer to it with entire confidence.

That Samoen Deschnew, in his celebrated voyage from the Colima or Kovima, to the Anadyr, in 1648, did not, in reality, double the north-east cape (Schelatzkey, or Swoetoy-noss, the great cape of the Tschukutskoi,) but crossed it by land, over a narrow isthmus, as was subsequently done by Staras Staduchin, appears to us an arbitrary assumption, which the accounts do not authorize, and which is sufficiently refuted by Deschnew's intention to build a ship at the mouth of the Anadyr, to send back the extorted tribute to Jakutzk, by the same way.

* A Memoir on the Geography of the North-Eastern part of Asia, and on the question whether Asia and America are contiguous, or are separated by the Sea. By Captain James Burney.—Philosophical Transactions, 1818, refuted in the Quarterly Review, June 1818.

A Chronological History of the North-Eastern Voyages of Discovery, by Captain James Burney, F.R.S. London, 1819.

Even if the documents, of which Müller, Coxe, and Pallas were in possession, and from which they communicated to us Deschnew's voyage, were lost, these men appear to us sufficient sureties, and we unhesitatingly receive their authority; that in this one case the north-east cape, or Schelatzkoy-noss, was doubled by sea.

Other accounts and traditions of a similar voyage appear to ourselves unauthorized. We readily credit the statement of Dauerkin, communicated to us by Sauer, that Schalauroff perished, in 1664, in the Icy Sea, and not at the mouth of the Anadyr; and we have no faith in the voyage of Laptiew, in 1740, as it is pretended to be related, according to Gmelin's verbal declarations in the Memoires et Observations Geographiques et Critiques sur la situation des pays septentrionaux. Lausanne, 1765, 4to. p. 42.

The European harpoons, found sticking in whales, by Hendrick Hamel, on the coast of Corea, in 1653, and again, in 1716, by Henry Busch, on the coast of Kamtschatka, appear to us of some moment. Burney, in opposition to Müller assumes that Busch merely repeated what Hamel saw, and this supposition appears to us very arbitrary. He is farther of opinion, that the Russians, long before the time of Busch, may have introduced the use of European harpoons on these coasts; but this we know, from our own knowledge, not to be the case. The Russians, weak in number in this part of the world, appropriate to themselves the fruits of the industry of the people whom they conquer, without bringing them new ones; and whales are pursued to this day by the natives of the Aleutian islands, only after the old manner, with their own harpoon. Every other confirmation of the fact would be useless.

We find, out of the limits of Burney's works, another fact, which Barrow has not mentioned in his Chronological History of Voyages to the Arctic Regions, Lond. 1818, and which appears to us to require some attention.

According to the accounts collected by Mackenzie, at the mouth of the river called after his name, a ship, or very large boat, manned with white men, visited this coast towards the year 1780, and the Esquimaux procured iron from it in exchange for skins. Mackenzie River seems to discharge itself into the sea, between two very far projecting tongues of land. The sea on the west, in which the ship appeared, has received from this circumstance the name of Belhoullaï Tou, White Man's Sea. It seems natural for us to imagine that this ship reached that spot through Beering's Straits.

There is indisputably a northern current in Beering's Strait itself, at least during the summer months. We found this current pretty strong on the 16th of August, on the Asiatic side of the strait. Its effects carried us considerably back, when we came out of the strait to double Cape

East; and, in this, our experience perfectly coincides with that of Cook and Clerke. But it is just at this season of the year, in which the melting snow on the shore must necessarily produce a southern current, if this sea formed a closed basin. As the streams in Switzerland, which descend from the glaciers of the Alps, swell and become more rapid in summer, the water in this basin must increase in the same season, from a similar cause, and flow out of its proportionably narrow and shallow entrance.

Other facts also prove the northern current in Beering's Strait. On the breaking up of the ice in the sea of Kamtschatka, the icebergs and fields of ice do not drift, as in the Atlantic, to the south, nor do they drive to the Aleutian islands, but into the strait to the north. On the 5th of July 1817, the ice had broken up on the southern coast of St. Lawrence Island, and we came there on the 10th, without having met with drifting ice. It was not till the night of the 11th, that we found this ice, as we approached the east point of the island towards the north. On this side of the island, the sea is not so deep, and the current less violent than on the Asiatic side.

It is to be observed, that in the Sea of Kamtschatka, the south winds prevail during the summer, and the north winds set in about September, and continue during the autumn. The influence of the winds on the currents cannot be denied.

The quantity of drift wood which the sea brings to the north, and casts up, and among which there are decidedly southern kinds of trees, as well as northern firs *; the seeds of well known southern siliquose plants, which are washed on the shores of Oonalashka, as well as of Radack, (though less numerous †,) cannot make us conclude with certainty, on a general motion of the waters of the Great Ocean, towards the north. On the one side, northern trees are thrown up at Radack, as well as southern ones at Oonalashka, and on the other side, as Beering's Strait decidedly offers too small an outlet for such a current, it seems to us more natural to suppose, according to theory, in case the fact was decided, that a double current takes place in the sea, as in the atmosphere. An upper one of the warmed lighter water towards the north, and an under one of the cold heavier water to the equator.

* We have seen at Oonalashka some joiners' work, in which only the drift wood thrown upon the shore of these islands was used, and which was distinguished for the great variety of beautiful species of wood. But the high north produces only birches and firs, and here, only far in the interior of the continent. We saw on the same island a large wrought block of camphire wood, which had likewise been cast up by the sea. The traces of the hand of man necessarily weakens it evidence. It might have been brought by some ship.

† They were eagerly sought by the Aleutians, as a particular superstition was attached to these *swimming stones*. They are said to be more numerously washed up on the eastern coast of the island.

The inhabitants of the Aleutian islands, of St. Lawrence Island, and of the shores of Beering's Straits, possess no other than drift wood. It is thrown up in different quantities in different years. It is to be observed that it is washed more to the American coast than to the Asiatic. In Kotzebue's Sound, we found it in large quantities, and, on the other hand, there was a scarcity of it in St. Lawrence Bay, where the Tschukutskoi burnt only moss and a little willow brush-wood. It might be asked, whether the accounts of forests on the opposite coast, might not refer as well to the drift-wood, in which they abound, as to the forests of Norton Sound, and the interior?

The alluvial sand hills on the American coast contain branches of trees and wood, of the same kind as that thrown up on the shore.

The drift wood of the north appears to us in general to be brought from the interior of the continent, by rivers and currents, and in the seas which we describe, to come particularly from America. Perhaps the river which flows into the sea between Bristol Bay and Norton Sound may be one of the most considerable sources.

The currents in the Icy Sea, along the coast of Siberia, are in general but little known, and we hesitate to draw conclusions from uncertain information. Liachoff and Schalauroff found the current to run west to the north of the Yana and the Colima, Sauer and Billing, in a west wind, east, and in a north-east wind, west. In Waigatz Strait, and to the north of Nova Zembla, the current appears to be likewise west.

After we have tried to prove that a current goes to the north, through Beering's Strait, we must confess, that it is too weak, and can force but too little water through the narrow entrance, to correspond with those currents which flow from Davis' Strait, and along the east coast of Greenland towards the south, as they are acknowledged to do during the season when these seas are open to navigation, and as many facts permit us to infer that they continue also in winter. *

The indications of land to the north of Beering's Straits, the flight of the birds from the north to the south, and the not increasing depth of the sea towards the north, from which Burney concludes the union of the two continents, seem to us sufficiently explained, by the supposition that islands, such as Liachoff's islands opposite the mouth of the Yana in the Icy Sea, might lie in this vicinity. The inhabited land of Andreef, or Andreanoff, to the north of the Colima, 1762, and the reports and traditions that it extended from the continent of America, as far as the new Siberia of Sannikoff, 1805, (the most eastern of Liachoff's islands,) appear to us equally unauthorized; and Burney himself lays no stress upon them.

* Quarterly Review, June 1818, p. 446.

We are therefore of opinion that the two continents are separated, and consider the north-east cape, or Schelatzkoy-noss, not to be an isthmus which unites both quarters of the globe, but merely a promontory of Asia, (like Cape Taimura between the Jenisei and the Lena, which was gone round and recognized only by Chariton Laptiew, in 1738, and that only by land.) The doubling of this promontory by sea has been prevented by the ice, and the examination of it by land, by the warlike independent people of the Tschukutskoi, since the time of Deschnew. Billing, whose instructions ordered him to solve this problem, by sea or by land, and who found all circumstances favourable, inexcusably neglected it.

We now turn to the north coast of America.

The North Cape of Cook, Mackenzie's River, the Copper-mine River of Hearne, are points that give us the principal direction, which runs nearly under the seventieth degree of north latitude. The accounts and charts of the Indians of Hudson's Bay, which all agree in continuing the coast of Copper-mine River northwards to Repulse Bay; the north-west current, and the same direction of the swell in Baffin's Bay, according to older authorities; the currents and tides in Roes Welcome; all circumstances concur to make us infer the connection of the sea and the separation of the land, and we seek the channel northwards of Repulse Bay, as far as Sir James Lancaster's Sound.* Captain John Ross, whose voyage confirms Baffin's former discoveries, affirms that he proved the connection of land in Baffin's Bay, against which, many persons who accompanied the same expedition, openly declared a contrary opinion, (the commander of the other ship, Lieutenant W. E. Parry, the learned Captain E. Sabine, the surgeon A. Fisher, &c.) and the often disputed question remains still undecided.† There remains, in every case, the coast from the entrance of Cumberland Strait to Repulse Bay still to be examined.

But whether the passage in the favourable season can be found open and free from ice, whether the north coast of America, in its whole extent, and with its promontories, if there are any such, can

* On the other side, whales, which have been harpooned near Spitsbergen, and which were found in the same season in Davis' Strait, as well as other circumstances, have given weight to the conjecture that Greenland is an island, or a group of islands.

† John Ross, Voyage of Discovery, &c. London, 1819.

The Quarterly Review, May 1819, p. 313., blames Ross for not having properly examined the promising Lancaster Sound. "There occur unfortunate moments in the history of a man's life, when he is himself unable to account for his actions, and the moment of putting about the Isabella would appear to be one of them." p. 351.

Modern Voyages and Travels. London, 1819. The Journal of Mr. Fisher.

Blackwood's Magazine, December 1818.

Captain E. Sabine. Journal of Literature, &c. April 1819.

The same. Remarks on the late Voyage of Discovery.

———. The Explanation of Captain Ross, &c.

be navigated as the Asiatic coast has been, in places, and at different times, is a question which we must leave undecided. The sea can be open but a few days in these high latitudes, and all circumstances combine to render discoveries more difficult, and to lessen their certainty. During the summer season a thick fog hangs over the sea, which only dissolves when it is driven by the wind over the warm land, and you do not behold at sea the sun which shines upon the coast.*

We observe, that that part of the American

* We observed this phenomenon particularly at St. Lawrence Island, at Oonalashka, in the bay of Avatschka, and at San Francisco.

The phenomenon of the parhelia, which is said to be often seen in the north of the Atlantic Ocean, is rare in the sea of Kamtschatka. We did not observe it ourselves, and a Russian who has grown old in the Aleutian islands never saw it more than once in his whole life.

We observed the phenomenon of the mirage, the most remarkable in Beering's Strait, and particularly at the entrance of Schischmareff Bay, where it surrounded us with manifold deceptions on shore, and on the sea, at all times of the day, like an enchantment. (Compare Capt. Ross's Voyage, p. 147.) The objects lying on the horizon seem to separate themselves from it, and to rise above it; (generally about three to five minutes, measured by a sextant,) they are reflected in the circle which is caused by their distance from the horizon, and seem to be prolonged by their reflected image. The causes of this phenomenon appear to us to have lain rather in localities than in the change of the atmosphere, and we have observed it in different zones with tolerable constancy, in the same place; for example, in the harbour of Hana-rura, (with the prospect towards the west,) in the Bay of Manilla, &c. but never in the vicinity of low islands.

coast which we examined to the north of Beering's Straits, appeared to us to excite the hope of finding a channel, among the entrances and friths which intersect it, which might lead to the Icy Sea, towards the mouth of the Mackenzie River, without doubling the Icy Cape, which would then be part of an island.* The already mentioned account of a ship in this sea, leads us even to conjecture that such a channel has been already navigated.

It now remains for us to discuss the last question.

Blocks of rock, which are frequently observed on floating ice-bergs of the north, and other indications, teach us that these ice-bergs were originally formed next the land, and it has been attempted to prove, by scientific reasoning and experiments, that ice cannot be formed, except in the contiguity of land, and that an open deep sea, without land or islands, cannot freeze, but must be found open and navigable at all times. We have to oppose only one fact against this notion, which, in our opinion, has been too little regarded; it is the state of the sea round the south pole, unless, by a very arbitrary supposition, to which nothing entitles us, we should represent the southern fields

* Several journals have published a letter from the author of these articles, (San Francisco, New California, on the 28th of Oct. 1816,) in which this opinion was delivered. An error of the copyist has altered the sense, so as to make it seem as if this entrance had really been examined by us.

of ice, as attached to an undiscovered, inaccessible continent. In its whole extent, land has been seen only in one point, namely, the Sandwich Land, and this, it is to be presumed, is like New Georgia, an island of inconsiderable extent, thrown up in the wide desert of the southern ocean.

We cannot attach any belief to an open north polar sea.

The mass of evidence collected by Barrington and Beaufoy*, though each might be separately disputed, appears to us to prove, incontestibly, that in favourable years, the sea to the north of Spitzbergen may be found entirely free from ice, and open for navigation to very high latitudes, as it really was found in 1754, 1773, and other years. But it equally proves, that in other years, and frequently, the ice has hindered, and will hinder the advance to the north, even under the eightieth degree of latitude.

If to the north of Scandinavia, between Spitzbergen and Nova Zembla, the sea is sometimes found open under very high latitudes, perhaps under the pole itself, while in other points, for instance, to the north of Beering's Strait, it is, perhaps, seldom found free from ice, even under the seventieth degree; if in the north of Europe the polar glacier, in which we believe, may be

* The possibility of approaching the North Pole asserted, by Barrington. A new edition, with an Appendix, by Beaufoy. London, 1818.

hollowed out by a deep inlet penetrating towards the pole, we are of opinion that these anomalies must be attributed to local circumstances, which cause the temperature, and, probably, to those that produce the much warmer climate, which that part of the world that we inhabit is known to enjoy, beyond all other countries of the northern hemisphere, under the same degree of latitude, which give woods and corn to Lapland, as far as to the seventieth degree, and support vegetation in Spitzbergen, even to the eightieth degree, and make that country able to maintain numerous herds of rein-deer, which the desert and naked Nova Zembla is unable to support.

At a time when men, like Humboldt, Buch, Wallenberg, &c. are labouring to increase the mass of experience, and a Humboldt attempts to look over, to investigate, and to bring under one law, the fragments of local meteorological observations which we now possess, as scanty contributions to physical geography; to draw isothermal lines over the globe; let us be permitted to submit to the examination of natural philosophers, an hypothesis for the explanation of the phenomenon.

We ask whether the theory which explains the sea and land breezes on the coasts, alternating by day and night, the local summer and winter monsoons, and, lastly, the general trade-winds, might not suffice in most cases, to explain the local differences of climate under the same latitude?

When we cast our eyes upon the globe, it appears to us that the twofold current of the atmosphere, from the equator to the poles, in its upper regions, and from the poles to the equator, in its lower region, determined in their direction by the rotation of the earth on its axis, must bring over Europe, from the interior of Africa, which is scorched by the sun, a current of air far more heated in proportion, than over any other country in the world. We consider the continent lying to the south and S.W. of Europe, between the line and the northern tropic, as a furnace which heats the air that passes over it, and determines its climate; a furnace such as no other part of the earth enjoys; and we think, in general, that continents lying between the equator and the tropics, must give to the more eastern parts of the world, towards the pole, a warmer climate than other parts have under the influence of similarly situated seas.

This is not the place farther to develop and discuss this idea, or to submit a new theory to calculations, and try it by the touch-stone of facts still imperfectly known. We only wished to point out the thought which struck us, passing travellers, in the north of the Great Ocean, at the sight of the wintry Aleutian islands, (under the latitude of Hamburg) and the coasts of Beering's Straits, (under the latitude of Drontheim and Norway. We shall now endeavour to bring these countries nearer to the reader's observation.

The points at which we stopped and endeavoured to investigate nature, are the following, from south to north:

The protected harbour of St. Peter and St. Paul, in the interior of the bay of Avatscha, on the east coast of Kamtschatka - -	53°	1′ N. Lat.
Oonalashka, one of the Fox islands, and in the line of the Aleutian islands, situated east of America	54	00
The island of St. George - - -	56	42
And the island of St. Paul, in the sea of Kamtschatka, north of Oonalashka - - - -	57	5
The South Cape, of the island of St. Lawrence, in the year 1817	62	47
And another part of the same island, in 1816 - - -	63	13
St. Lawrence Bay, on the Asiatic coast, to the back of which we penetrated by land - - -	65	35
The entrance of Schischmareff Bay, on the American coast -	66	13
The Rocky Island, in the interior of Kotzebue's Sound - -	66	13

and several points, lying a few minutes farther to the north, on the shores of this sound.

We have beheld at St. Peter and St. Paul, from the

29th of June to the 13th of July, 1816, the first dawn of spring. The year was backward; the early *Anemones* and *Corydales* were only in their first blossom; the snow melted on the hills, well covered with vegetation, which entirely surround the harbour, and they gradually assumed their verdure. At the time of our departure, the first roses were blowing, the first blossoms of the rhododendron, the lilies, &c. and the snow still crowned the mountains and covered the bases of the high volcanic pyramids, which tower above the land, and have been trigonometrically measured by the indefatigable Hörner. The season was unfavourable; and we do not flatter ourselves with being able to extend the scanty knowledge which we have of the nature of this country. We refer to Krascheninikoff, Pallas, Steller (Description of Kamtschatka, Frankfort, 1774), Lesseps, and other travellers. Krusenstern affords, in other respects, complete information, in his account of Kamtschatka.

The bay of Avatscha lies between the latitude of Berlin and Hamburg; and the harbour of St. Peter and St. Paul, in the interior of it, seems to be as little exposed to the influence of the sea-wind, as the interior of the friths of Nordland. The birch still grows here in the form of a tree, but stunted and unlike the slender beautiful trees which we admire in the north of Europe, and particularly about St. Petersburg. *Pinus Cembra,* which, on our Alps, grows higher than the *Pinus Abies*, and marks the limits of trees, *Sorbus aucuparia*, *Alnus incana*, and several willows, retain the form of shrubs. Timber is obtained from the interior of the peninsula, which enjoys a milder climate than the east coast; and the seeds of the *Pinus Cembra*, which we meet with on the tables of the Russians, come from Siberia, by way of Ockotzk.

Grasses and herbs grow with great luxuriance, on a rich soil, in a damp atmosphere. There are but a few kinds of plants, and they are every where equally distributed. In shady places grow *Spiræa Kamtschatica*, *Allium ursinum*, *Mayanthemum canadense*, *Uvularia amplexifolia*, a *Trillium obovatum*, (of Pursch), &c. On the plains, a *Veratum*, *Lilium Kamtschaticum*, *Iris sibirica*, &c. On rocky hills *Caprifolia*, *Spiraceæ*, roses, *Atragene alpina* and alpine plants, such as *Rhododendron Kamtschaticum*, *Empetrum nigrum*, *Trientalis Europæa*, *Linnea borealis*, *Cornus suecica*, *Saxifragæ*, &c. Several ferns, by their number, constitute a considerable part of the vegetation. There are some *Orchideæ*. *Urtica dioica* is evidently imported, but become naturalized.

We are of opinion that summer corn would thrive in St. Peter and St. Paul, as in Lapland, under the seventieth degree, and in the valleys of the Savoy Alps, (*au Tour*, &c.) In its absence, the potatoe grows tolerably, though it produces but small roots, and this root, which supplies the greatest part of Europe in the place of corn, might be here of the greatest importance. Brandy might be distilled from it, and provide for the chief want of this colony. But there is still a greater want of hands and industry than of productions and productive power in the earth, and even what has once been undertaken with advantage, as, for instance, the boiling of salt, is neglected. Krusenstern justly observes, that the earth is cultivated too late. The mountain of transition slate, which separates the harbour from the bay of Avatscha, has slate which would conveniently supply the town with stone for building; and lime might be burned out of shells, if lime-stone cannot be found.

Innumerable smoking volcanoes rise along the ridge, which, extending between two continents, in the form of an arch, constitute the chain of the Aleutian islands, and tower, in pyramidical forms, above the clouds. Torn, rugged rocks, form, in broken lines, the ridge which unites these threatening colossuses. The ridge appears to decline, from the American continent, over the peninsula of Alashka; and the chain of islands towards Asia. These islands become of less extent towards the west, and less frequent, and the last of them, Beering's Island, inclines in a gentle slope towards the Kamtschatka coast.

The two peaks of the peninsula of Alashka are of extraordinary height. The first, in the north-east, which fell in some years ago during an eruption, still appears, with its truncated summit, to be the highest. The next following, a sharp-pointed cone, is evidently considerably higher than the peak of Oonemak, and this, which seems to exceed the mountain Mackuschkin, in Oonalashka, and the similar summits of the nearest island, is 1175 toises high, according to the measurement of M. Von Kotzebue. The snow entirely covers the cone, and its basis, after a rough estimate, on the two upper thirds of its height, and descends down still farther to the shore in many places.

The sight of this mountain leaves an extraordinary impression behind. The eye which has accustomed itself, in our Alps, to use the line of snow as a rough approximation, can with difficulty, withstand the deception of over-rating the heights of these summits.* The line of snow which Wahlenberg has observed in the Swiss Alps, at 1371 toises, and in the Lapland mountains, at 555 toises, and Leopold Von Buch, at Mageroe, 71° north latitude, at 333 toises, might, according to our rough estimate, descend over these islands to about 400 or 300 toises; and separated summits, which do not attain this height, are covered with snow on their tops, and in clefts and openings of their

* From the same cause there was an opposite effect in Teneriffe. The peak, which was scarcely touched by the snow when we saw it, did not make the impression on us which its great height gave us reason to expect.

sides. In the autumn of 1817 the snow had remained in many places where it had vanished in the autumn of 1816. The sources in the low valleys of Oonalashka, which we examined towards the beginning of July 1817, showed us the temperature of the earth between 38° and 39° of Fahrenheit.*

Granite is found at Oonalashka. The mountains of the interior, to the left of the valley, which goes from the chief settlement to Makuschkin, are of granite. Otherwise, on all the shores of the Great Bay, on the road to Makuschkin, and even at Makuschkin itself, we found only clay porphyry, passing on the one hand, and principally, into amygdaloid, and on the other hand into greenstone, imperfect conglomerate porphyry, and real conglomerate.

These kinds of rock lie over one another in immense strata, not much inclined, and alternating apparently without any law. The stratification is to be seen only at a distance, in the profile of the mountains. These porphyries offer only large sharp-cornered, pointed, acicular shapes, and only where

* We regret that the state of our meteorological instruments, of which we had already lost several, and to preserve the last of which was our duty, hindered us from repeating our observations, and bringing our results to a satisfactory conclusion; for this cause we were unable to use the barometer for taking the altitudes.

they conglomerate, rounded forms, as granite frequently does.*

Out of several places in these porphyry mountains break hot springs, the water of which is tasteless and without smell, and deposits upon the stones a slight covering of yellowish-brown stalactite. Dr. Eschscholtz found the temperature of one of these springs, which rises in a meadow in a valley, opposite to the entrance of the harbour, between 93° and 94° Fahrenheit. The stagnant water of several brooks, in the same meadow, deposits a bright yellow sulphureous sediment. The water of the spring already mentioned, and another on the island of Akutan, in which meat was thoroughly boiled in a short time, seemed to the Doctor to be distinguished by a greater quantity of lime in the water, than in common springs. Another hot spring flows from a layer of real conglomerate, near Makuschkin at the foot of an insular detached hill, of inconsiderable height, on the sea-beach, below high-water mark. The strata lying on it, of which the hill consists, afford the usual varieties of clay porphyry.

The Makuschkaia-sobka continues to smoke without any eruption, and the Aleutians procure sulphur out of it. We did not penetrate to the

* We are indebted to Professor Weisz for most of the geognostical remarks occuring in these essays, and who has had the kindness to examine with the author all the specimens of rock.

separate enormous mass of rock which bears this fiery crater; and in the two parts of the island through which we traversed met with no real lava.

Iron pyrites deceived the rapacity of the first discoverers at Oonalashka, as in so many parts of the world, who mistook it for gold.

At Oonalashka we procured petrified wood, fragments of large dicotyledon stems, which were said to have been procured from the bed of a lake in Umnack, which had dried up in consequence of an earthquake. The volcanoes of this island are particularly active, and the stones thrown up by them have in later times filled up a channel which was formerly navigable.

The new island which rose from the waves in the year 1795, in the neighbourhood of Umnack and Oonalashka, and of the origin of which Langsdorff gives an account, is said to be already covering with vegetation.

Trees are still found on the peninsula of Alashka, and on the neighbouring island of Oonemak, which is separated from it by a narrow channel, and which seems to partake of the nature of the continent. Oonalashka, and the other islands of the chain, are entirely destitute of them. It has been attempted to plant pines, a kind of *Abies*, brought from Sitka, at Oonalashka; most of them have perished, the others seem scarcely to thrive, but the plantation is still young, and it is well known how ill coniferous trees bear transplanting.

At Oonalashka, where we resided at three different times, in spring and autumn, we were particularly engaged in studying its Flora; and this island will serve as a point of comparison for the other more northern countries which we visited.

At Oonalashka (under the latitude of Lubeck), the willows in damp places scarcely rise above the luxuriant growth of grass and herbs. Immediately you ascend from these plains to the next hills, you find an entirely alpine Flora, and only some kinds of *Vaccinium myrtillus* grow as shrubs in the lowest region of the mountains. Otherwise a damp atmosphere nourishes the verdant mantle of the earth, up to the more naked rocky summits and to the shining snow; and several fine plants adorn this gloomy world with an admirable splendour of colour. (*Lupinus nootkaensis, Mimulus luteus,* Pursch. *Guttatus,* Willd. En. Sup., *Epilobium angustifolium et latifolium, Rhododendron Kamtschaticum,* &c.) The fresh verdure of the meadows reminds us of the valley of Urseren.

The Flora seems to have no other affinity to that of St. Peter and St. Paul, than that which it derives from the general alpine or arctic Flora, and the strand Flora of these northern coasts. Besides such plants as we again met with in higher northern latitudes, we observed at both places only the *Lilium Kamtschaticum* (if the variety at Oonalashka is not a peculiar species) and the *Uvularia amplexifolia;* and, on the contrary, found on the

American coast, to the north of Beering's Strait, several species of Kamtschatka plants, which we did not see at Oonalashka. It is the Flora of the north-west coast of America, which extends to the foot of the hills of this island, where it unites with that of the arctic regions.

We mention as examples, *Rubus spectabilis, Lupinus Nootkaensis* (which, though stunted, grows up the mountains), *Epilobium luteum*, and *Mimulus guttatus*, Willd. * The *Claytonia Unalaschensis*, Fish. Siberica hort. (*Alcinoides*, Pursch. ?) might perhaps be also enumerated here. *Sanguisorba canadensis, Lithosperam angustifolium*, &c.; belong to the general Flora of America.

Many kinds of grass flourish in the low places; with them several *Umbellatæ, Angelica, Heracleum*, &c.; a dozen of *Carices* scarcely make a more considerable part of the vegetation than in the north of Germany: several *Scirpus* and *Eriophorum* accompany them; the *Junci* combine with them in a proportion of about one to two. The *Orchideæ* maintain a considerable rank, as well by the number of their species, as of the individuals, in the Flora of the vallies and the mountains; we counted eleven of them, among which a beautiful *Cypripedium* was the most distinguished. We did not observe a single plant of this family higher to

* The seed of this plant, which is cultivated in the botanic garden at Berlin is said to come from Lake Baikal (?)

the north. There are about eight species of fern; farther north we met with only one *Filix*, and this no more than once. Several *Lycopodia* are found as far north as Oonalashka; more northerly only one species. Several water-plants are found in the lakes: *Potamogeton, Sparganium, Ranunculus aquatilis*, &c. Farther north, we observed only the two species of *Hippuris* and the common *Callitriche*.

Two other ranunculus, the *Prunella vulgaris*, a *Rhinanthus*, a *Cineraria*, an *Achillea*, *Plantago*, a *Geum*, several *Rubiaceæ*, a *Claytonia*, the *Menyanthis trifoliata*, a *Triglochin*, &c. belong with the above-mentioned plants to the Flora of the vallies of Oonalashka. A *Bartsia* seems to differ from the *Bartsia pallida* which is found farther to the north. A beautiful plant, which forms a new and distinguished species, the *Romanzoffia Unalaschkensis*, received the name of the patron of all the sciences in Russia. The species *Rumex*, *Polygonum*, *Aconitum*, *Talictrum*, several *Alsinaceæ*, the *Iris sibirica*, the *Geranium pratense*, the *Comarum palustre*, the *Montia fontana*, are spread over the whole north.

The *Empetrum nigrum*, which with *Helleborus trifolius*, Lin. (an American plant which we did not meet with farther north) for the most part clothes the hills, opens the kingdom of the Alpine Flora. There are several species of *Vaccinium* and the *Oxycoccos*, *Arbutus alpinus*, and *Uva ursi*, a white-blossomed *Menziesia*, which is included with the *Erica cærulea*. *Rhododendron Kamtschaticum*, *Azalea procumbens*, *Andromeda lycopodioides*, which is replaced higher to the north by the *Andromeda tetragona*, Alpine *Salices*, *Sylene acaulis*, *Sibbaldia procumbens*, *Cornus succica*, *Trientalis Europæa*, *Linnea borealis*, *Ornithogalum striatum* *,) *Anthericum calyculatum*, L. *variet. borealis*, *Königia islandica*, a *Gymnandra*, which seems different from that found more to the north, ten *Saxifragæ*, three *Pediculares*, several *Potentillæ*, two *Gea*, two *Anemonæ*, three *Primulæ*, a *Papaver*, a *Drosera*, a *Pinguicula*, two *Pyrolæ*, one *Viola*, one *Parnassia*, one *Rubus*, and one *Armeria*. Only one Alpine *Ranunculus* and three *Gentianæ* are met with, of which species several kinds are found more to the north. Of the class of the *Syngenesiæ* the *Aster*, *Hieracium*, *Gnaphalium*, *Leontodon*, *Arthemisia*, &c. are met with. This class is more numerous higher north, where the genus *Arthemisia* has several distinguished species. On the other hand, several Alpine species of the genera *Campanula* and *Veronica* are met with at Oonalashka, which are entirely missed farther north. Several kinds of the class of *Satyriun* are strewed partly in the valley, and partly on the mountains.

At Oonalashka we missed *Alnus incana*, *Betula nana*, *Ledum palustre*, *Dryas octopetala*, *Diapensia lapponica*, *Rhodiola rosea*, the genera *Spiræa*,

* Two varieties of this plant may, perhaps, be different species.

Astragalus, *Allium myosotis*, *Corydalis*, *Valeriana*, *Aretia*, *Androsace*, *Dodecatheon*, *Delphinium*, and *Orobanche*, which we found farther to the north.

The strand Flora, which farther north remains the same, without alteration, is formed particularly by the *Elymus mollis*, herb. Görenk. *Trinius* in Sprengel's Ent. ii., p. 72. *Arenaria peploides*, *Pisum maritimum*, several varieties of the *Pulmonaria maritima*, Willd., which are perhaps peculiar species, *parviflora*, Pursch. *Cochlearia officinalis* and *Arnica maritima*, which grows luxuriantly, and branching on this island, but produces only one flower in higher latitudes. We might add to this Flora the *Potentilla anserina*.

The sea along the coasts, and in the bays, is rich in *Algæ* and the *Fucus esculentus*, the sea-cabbage of the Russians settled there, is distinguished among many species of gigantic Fucus.

The mosses and lichens begin already at Oonalashka to occupy that great space in the Flora which they maintain further northward.

The island of St. George, with flatted ridges of fragments of rock, and steep shores, forms a table of tolerable height, and inconsiderable extent, to which a plain joins on the east side. You perceive the strata in the profiles of the shore: the kind of rock appears to be clay porphyry, as at Oonalashka, and large blocks of porous lava, partly form the strand.

The island of St. Paul is of greater extent, and

lower than St. George. Only small hills rise in the interior, one of which is in the form of a very truncated cone. The shores gently slope towards the sea, and form several capes and peninsulas. Some reefs extend from the island and from a rock (the Bober Island), into the sea, and are not without danger for ships. The peninsula, on which the settlement lies, is formed partly of heaped volcanic scoriæ, and partly of a porous lava, resembling scoriæ of iron; the irregular surface of which, not being yet grown over in several places, sets it beyond all doubt, that it was once fluid. Has this current, (of lava), risen from the sea, or has it been thrown out by a mountain, which fell into itself? It can hardly have rolled on, in the present state of the island, from the distant and low hills of the interior, on an almost horizontal plane to the shores. A profile, near the landing-place, clearly shows horizontal strata.

Fire has been several times seen from St. George and St. Paul, burning at sea; and people have imagined, in clear weather, that they could distinguish land to the south-west of St. Paul. Our examinations have proved that this latter appearance was only a deception: the fire might have been volcanic.

We had only a cursory view of these islands, which lie nearly in the latitude of Riga; it is remarkable how much more wintry nature appears on them, than at Oonalashka. Covered valleys and plains do not, as there, cherish a more luxuriant vegetation, and more southern plants. An entirely Alpine Flora unites itself, as further north, to the strand Flora. The raised ridges of mountain fragments are overgrown with black and pale-coloured lichens, the places watered by the melting ice with *Sphagnum*, mosses, and a few *Carices.* The earth has no more springs: the different arctic plants choose, according to their nature, rocks or moorland, and none of them rise above the soil on which they creep. The *Lupinus* at St. George, the *Achillea* at St. Paul, remind us of Oonalashka; but several plants, which are not met with at Oonalashka, of the higher latitudes. *Ranunculus, Pallasii,* and *Gmelini,* an *Androsace,* a *Claytonia,* &c. We found a single plant (a *Cochlearia?*) exclusively on these islands, where it is frequent and characteristic.

Both islands, before they were discovered by the Russians, were unknown to the neighbouring people; and the peaceable abode of the water-fowl and seals (*Phoca leonina* and *ursina*). On both of them, there are only Aleutians under the superintendance of the Russians; and the animals are subdued as well as the people. The island of St. Matwey, (Choris Island), is still uninhabited; the fate of the intended settlement is known. During the winter, the people found themselves abandoned by the animals, on which they had reckoned for their support; and starved, with the exception of three, who prolonged their lives with a poor clay, which they discovered. We received a specimen of this mineral at Oonalashka, which had been brought there by previous navigators; it is in the European collections.

Islands like these, and so near to inhabited coasts, would not have remained uninhabited had they been in the Great Ocean.

We shall unite St. Lawrence Island, and the two shores of Beering's Strait, in one point of view. They are formed of the same primitive mountains, and the same Flora is spread over them. These countries lie between the latitude of Christian Sound, and Dönna öe on the Norwegian coast, or Hernosand and Tornea on the Swedish, in the Gulf of Bothenia.

St. Lawrence Bay is a frith of the Asiatic coast, which penetrates into the mountains; the background of which is formed by eminences with naked rocky declivities. The kind of rock is primitive limestone.

The Alpine or arctic Flora, which here adorns the foot of the mountain, does not seem to crown their summits, even when free from snow; and while the clefts where the snow has melted are adorned with the richest vegetation, the dry ridges and declivities of piled masses of rock are scarcely tinged with grey and black lichen.

The mountains in this wintry climate, uncovered and unprotected by vegetation, moulder and decay.

The frost bursts the rocks; the mild heat of every summer brings down new masses, and the destruction continues, till it is complete. The soil is every where composed of fragments of rocks, except where the *Sphagnum* has formed a turf, or marshy soil, in deep watered places.

The poet of a happier climate, shows us, in the wood-covered summits of his mountains, the image of indestructibility, while the gloomy song of the northern bard exemplifies by his rocks the destructive power of age.

The Tschukutskoi, who inhabit St. Lawrence Bay, possess a tolerably large quantity of a fine graphite, with which they paint their faces with crosses and different figures, as an ornament. We have obtained, of the different tribes with whom we traded of both coasts, nephrites wrought in different manners, on which, with their present riches in iron, they place no particular value. We have been unable to discover where those two minerals are found.

St. Lawrence Island is of moderate height, and its ridges are flattened. At the place where we landed, in the year 1816, we met with a kind of rock mixed with green-stone; and in the year 1817, to the east, and near the south cape, we also found, as it were at its foot, large masses of granite. The forms which, when viewed from the sea, the profile of this promontory exhibited, excited our curiosity, and we believed we had seen in them basaltic columns, which leaned, nearly perpendicular, in one direction to the south.

The American coast, to the north of the strait, is formed of alluvial levels and downs, between Cape Prince of Wales and Cape Krusenstern, (Cape Mulgrave, Cook?) which are two rocky pillars. Kotzebue's Sound led us through them, as far as the primitive land, to which they are annexed. The land is raised but a little, and the uninterrupted lines of the hill prevent us from seeing where the rocky ground begins.

The rocky island which protects the anchoring-place in the back-ground of the sound is of a mixed kind of rock (flinty slate). It has a powerful influence on the magnetic needle, and changes its direction. The rock shows itself again on the profile of the opposite shore, which forms the bottom of the sound. Eschscholtz Bay, into which the sound is prolonged to the north-east, penetrates again into alluvial land. We landed at a sandy point on the east side of the bay, where the magnetic needle was likewise found to vary in an extraordinary manner. Shall we from this anomaly infer the vicinity of primitive mountains which we do not immediately see?

Docter Eschscholtz intended to return along the strand of this sand-bank to the rocky shore, which is a continuation of it. He found between the sand and the primitive mountains, for which he was looking, an imperceptible continuation of both, without being able clearly to recognize the stratification, a kind of rock, which, as far as our knowledge extends, only Link has reckoned among rock, viz. ice, pure solid ice.

The profile, where, being indented by the sea, it comes to sight, reaches to the height of eighty feet at the most, and the highest ridge of the hill scarcely double. Over the ice is a covering of blueish clay (or loam), from two to three inches thick, and immediately over that a kind of turf scarcely a foot deep. The vegetation is perfectly the same as on the alluvial sand and clay soils. The earth thaws only a few inches deep everywhere, and it is not to be ascertained, by digging, what soil it is. The mould, which falls down from the ice-hills, protects their foot, and a stop is put to farther devastation when under this falling earth a declivity is formed, which reaches from the foot to the summit. The length of the profile, in which the ice is exposed to sight, may be about a musket-shot. But it is evident, in the forms of the overgrown declivities of the shore, that the same kind of mountain (ice) occupies a much greater extent.

We are already acquainted, from several travels, with similar ice-ground in the north of Asia and America, and among these, particularly, the rocks of ice, covered with vegetation at the mouth of the Lena, out of which the mammoth, the skeleton of which is now in St. Petersburg, was thawed, and on which a cross was erected by Adams, to whom we are indebted for the preservation of the skeleton, and the accounts respecting it.

Fossil ivory is found here as in Northern Asia, and the natives manufacture utensils out of it as of morse and physeter teeth. We found, near the ice ground on the point of land where we bivouacked, and where the natives had stayed before us, some grinders, which perfectly resembled those of the mammoth, and also a tusk which, by its greater thickness at the root, and its simple curvature considerably differed from the well known mammoth's horns, and seemed to have much more resemblance to the teeth of the present race of elephants. During the night our watch-fire was partly kept up with such ivory.

We have remarked the greater riches of the arctic Flora, amidst manifold variety of soil on the rocky coast of St. Lawrence Bay; the greater poverty, on the other hand, on the flat sandy coast of America, whose hills are uniformly clothed with *Sphagnum*, and where only, on the rocky island in the interior of the sound, there are some species of Alpine plants which thrive only on a rocky soil. We collected many kinds of plants in St. Lawrence Bay, which we met with no where else. The equally rocky island of St. Lawrence, on which we stopped for but a few moments, showed us several kinds, which it has in common with the bay of the same name, and which are wanting on the American coast. Lastly, this coast offered us but a few kinds which we did not find in St. Lawrence Bay. We cannot give any more essential difference in the Flora of the two coasts, than that caused by the difference of the soil and climate.

The aspect of nature is the most wintry in the Bay of St. Lawrence. The vegetation, pressed down to the soil, scarcely rises perceptibly in its back ground, where the shrubby willows hardly reach the knee. The *Andromeda polifolia* which we found only there, was no more than two or three inches high, and bearing one flower. The Flora of this bay is adorned by a *Delphinium*, a *Dodecatheon*, an *Aretia*, and several species, which we saw only there, of every genuine arctic Alpine genus. *Gentiana, Saxifraga, Astragalus, Arthemisia, Draba, Ranunculus, Claytonia*, &c. Several of them had not been described.

St. Lawrence Island, lying two degrees farther south, does not differ from St. Lawrence Bay, in respect to vegetation. An *Andromeda tetragona*, the *Dryas octopetala*, the *Diapensia lapponica*, species of Alpine *Myosotis*, a *Gymnandra*, &c. mark, as in St. Lawrence Bay, the character of the Flora. We observe that in this island, transported into the arctic vegetable kingdom, we gathered, in a few minutes, more flowers in blossom than we had observed during several weeks on the island-chain of Radack lying between the tropics. Farther to the north, on the Rocky Island, in the interior of Kotzebue's Sound, grows the *Azalea procumbens*, as at Oonalashka, and in the bay and island of St. Lawrence; with this are Alpine willows, *Cornus suecica*, *Linnea borealis*, arctic species of *Rubus*, *&c.* *Empetrum nigrum*, and *Ledum palustre*, are every where met with on moor-land, and under the *Sphagnum*; but the *Ledum* does not form there the high bush which adorns the moor-lands in northern Germany.

The vegetation in the interior of Kotzebue's Sound is considerably higher than in the interior of St. Lawrence Bay. The willows are higher, the grass kinds richer, all vegetation more juicy and stronger. Most of the species of plants found by us on the American coast, and wanting in St. Lawrence Bay, indicate a less wintry climate. We found on the island mentioned above, *Alnus incana*, as a little shrub, and *Spiræa chamædrifolia*, plants which we had noticed at Kamtschatka, and not on the American island of Oonalashka, and which a more rigorous clime seems to have kept from St. Lawrence Bay. The Flora of this island is adorned by *Orobanche* and a *Pinguicula*. The *Cineraria palustris* grows particularly luxuriant on the well-watered declivities, which are formed at the foot of the ice-walls. *Betula nana* is met with already on the outer coast. The level ground of this coast is not covered with snow during the summer.

Not far from the back of Kotzebue's Sound, about a degree and a half more south, Cook found the banks of Norton Sound well wooded, and the trees rose always higher towards the interior of the country (northwards).

Mackenzie, farther east, in the interior of America, found the banks of the river to which he gave his name, grown with high trees, as far as under the 68th degree of north latitude, and these banks appeared to him to be of ice.

It seems to us, weighing all the circumstances, that the American coast of Beering's Strait enjoys a milder climate than the Asiatic.

Let us be permitted to place by the side of this melancholy portrait of these coasts, the picture of European nature under the 70th degree of north latitude (three degrees and a half further northward than the most northern points touched by us.) "The circular bay and the amphitheatre of Talvig, as they suddenly burst upon us through the narrow channel by which we passed, were highly attractive. The church stood in the middle of the green and animated slope, with the large house of the clergyman above, and on the sides there were two considerable gaards, with Quäns and peasants along the banks; and picturesque rocks, with a majestic foaming waterfall above. Add to all this, the animation of summer; the ships in the harbour, a Copenhagen and a Flensburg brig, with a Russian vessel from the coast of Archangel, and Finns and Norwegians in continual motion backwards and forwards in the bay, going with fresh fish to the Russians, and dry to the merchant, and returning with meal and grain. Who would represent Finmark as dreary and miserable, if he saw the beautiful situation of the Bay of Talvig?

"About mid-day we crossed the nine short English miles from Talvig to Altengaard, the seat of the head magistrate (*Amtman*), in the inmost part of the Fiord. This gaard is also an object of surprise. It is situated in the middle of a wood of high Scotch firs, in a green meadow, with noble views through the trees, of the Fiord, of the points which project beyond one another into the water, and, finally, of the fields of Seyland and Langfiord. The trees around are so beautiful, and so diversified! We see through the boughs, on the opposite side of the water, the foaming stream which descends from the rocks, and communicates perpetual motion to the saw-mills; and in the Fiord, and in Refs-botn, every hour that the sun advances in its course, lights up some new gaard to us. The habitation is a villa, not a country-house built for the dust of law papers, or for the management of law-suits. It appears, when we enter the wood from the beach, as if we were transported to the park of Berlin; and when the perspectives down the Fiord open on us, it then seems as if we were viewing Italian distances, or one of the lakes of Switzerland." (Leopold Von Buch's Reise durch Norwegen und Lappland, &c. p. 485.)

Mageroe, under the 70th degree, with its jagged, naked rocks, among which large and extended masses of snow lie every where, at the end of July seems to be a counterpart to the aspect of the shore of the Bay of St. Lawrence. The birch, however, though stunted, grows more on the declivities of the mountains, at an elevation of four hundred feet. Leopold Von Buch estimates the mean temperature of the air of this island at $1\frac{1}{2}°$ R., and the height of the perpetual snow at two thousand feet. It, however, never freezes in well-closed cellars, and the grass does not cease growing under the snow. A rivulet flows during the whole winter, near Hammerfest, at Qualoe.

We see, on the other hand, on the coasts to which our attention is directed, a more luxuriant vegetation; bushes, (high trees, Mackenzie,) thriving on a soil eternally frozen on a ground of pure ice.

Wahlemberg (de Vegetatione et Climate in Helvetia Septentrionali, p. 84.) has proposed this law for Europe. "The mean temperature of the air towards the 46th degree of north latitude, is equal to the temperature of the earth, in level countries, raised but little above the surface of the sea. From this centre, the temperature of the air decreases more quickly, as well towards the north as towards the summits of the mountains, than the temperature of the earth; and, towards the south, more quickly increases, so that in the north, and on the mountains, the temperature of the earth is warmer, but in the south less warm than the mean temperature of the air."

On the coast we visited, only the direct heat of the sun, and the temperature of the air, during the summer, could support the vegetation on an eternally-frozen earth. Shall we suppose the cold of the winter to be so severe there, that the mean temperature of the air could still fall below the temperature of the earth? The sight of nature on these coasts, in the absence of all meteorological observation, contradicts the above law, which, if proved for Europe, seems to be unfavourable to the hypothesis ventured by us, according to which the mild climate of this part of the world is owing to the warmer air which passes over it.

Steller, whom Pallas calls immortal, first developed, under Beering, the natural history of this country and these seas; and Merk, under Billing, honourably followed his example. Other learned men and collectors have explored Kamtschatka more at their ease, and Oonalashka has been visited. The names of Steller and Merk have retained all their lustre. Of their botanical collections much has remained unpublished, particularly in the herbaries of Lamberti, Willdenow, and Görenki. Pallas, in the Zoographia Rossica, as far as it went, (to the middle of the fishes,) has collected every thing relating to zoology. With due respect to our predecessors, we shall make but a few observations on the Fauna of these seas and coasts.

The large mammalia have gone over from the American continent to Oonemak. There we find the rein-deer, a wolf, and a bear, which appears to be the European brown bear. The black bear (*Ursus Americanus, gala genisque ferrugineis*), the valuable skin of which is sought for furs, is first met, together with the brown bear, on the remote north-west coast. There is, besides, at Oonalashka, the black fox, and several small *Glires*, among which the *Mus oeconomus* is distinguished, which stores under the snow, for winter stock, the roots of the *Polygonum viviparum*, of the *Surana*, (*Silium Kamtschaticum*,) and other plants. The other mammalia belong to the Fauna of the sea.

As, on the one hand, in proportion as you go further in the land towards the north, the woods become less lofty, the vegetation gradually decreases, animals become scarcer, and, lastly, (as at Nova Zembla,) the rein-deer and the *Glires* vanish with the last plants, and only birds of prey prowl about the icy streams for their food; so, on the other hand, the sea becomes more and more peopled. The *Algæ*, gigantic species of *Tang*, form inundated woods round the rocky coasts, such as are not met with in the torrid zone.* But the

* The sea *Tang*, which serves the Manilla galleons as a mark of the vicinity of land on the coast of California, might,

waters swarm with animal life, though all aquatic animals seem to remain in a lower scale than their relatives of the same class on land. The *Medusæ* and *Zoophytes*, *Moluscæ* and *Crustaceæ*, innumerable species of fish, in incredibly crowded shoals, the gigantic swimming mammalia, whales, physeters, dolphins, morse and seals, fill the sea and its strand, and countless flights of water-fowls rock themselves on the bosom of the ocean, and, in the twilight, resemble floating islands.

The sea-otter does not seem to penetrate to the northward beyond the chain of the Aleutian islands, and begins to become scarce after it has caused the destruction of the native tribes. The sea-lion and the sea-bear appear to keep in about the same limits; other seals, more resembling the *Phoca vitulina*, are found more frequently to the north. Countless herds of morse are met with in Beering's Straits, and the teeth of these animals seem to form a considerable branch of trade with the natives of St. Lawrence Island. We heard only corrupted traditions at Oonalashka, which seemed to refer to the *Manatus borealis*. A *Physeter*, a sea-wolf, six different species of whale, *Delphinus orca*, and two other dolphins, are found round the Aleutian islands; and, besides this, the

perhaps, mark the extreme progress of this formation to the limits of the monsoon. To this is added the *Fucus buccinalis*, which occurs at the Cape of Good Hope.

Delphinus leucas, in the north of Beering's Straits, as we infer from several accounts.

On the coast of Beering's Straits are found several species of *Viverra* and *Canis*, among which the black fox has chiefly excited our rapacity. The very common *Arctomys cytillus*, the skin of which produces an elegant fur, is distinguished among the *Glires*. The rein-deer, which belongs to both coasts, seems to be wanting in St. Lawrence Island. The dog, every where in the north the first companion to man, and his useful draught animal, is wanting only on the Aleutian islands, where it was formerly introduced and increased, but was extirpated by the masters of the country, because it pursued the fox, whose skin was their surest source of riches.

Many land-birds have spread over to Oonalashka from the nearest coast, of which the white-headed American eagle is predominant. With respect to the albatrosses, *Diomedea exulans*, we have to correct a very common error, which has gained credit under the authority of Pallas.* The albatross does not visit the north as a transitory guest from the southern hemisphere, merely to appease its hunger for a short time, and then to return at the

* Unica Septentrionum visitans avis Diomedea albatrus, hiemem antarcticam fugiens, per immensum oceanum ad nostra littora, aestiva abundantia piscium anadromorum allicitur, nec tamen apud nos generat, sed ad æstatem antarcticam prolificandi gratia illuc denuo abit. Zoogr. Ross, vol. i. p. 297, and vol. ii. p. 308.

breeding season to its southern home. The albatross builds its nest of feathers on the highest summits of the Aleutian islands, namely, on Umnack and Tschatirech Sobpotschnie ostroff, (the Island of the Four Peaks.) It lays two very large eggs, of a blueish colour, and hatches in the summer season. The black variety mentioned by authors is the young one. The Aleutians ascend these summits towards August, and take the eggs from the nest; they also throw darts, made for the purpose, at the sitting birds, and are particularly eager after their fat, with which they abound at this season.

Not a single animal of the class of *Amphibiæ* appears either at Oonalashka or the Aleutian islands.

Among the insects, the beetle is predominant; and among these the *Carabus*, of which Dr. Eschscholtz counted sixteen kinds, many of which are not hitherto described. Several water-beetles animate the lakes and standing water. It might perhaps be in vain to look for them more to the north.

The common northern large Maja (*Lithodes arctica*, Lat.) is distinguished among the crabs, and is particularly excellent for food.

We refer to Pallas and other authors with respect to the fish, on the constant and innumerable swarms of which the subsistence of the people of the north, and of their domestic animals *, (with the exception of the rein-deer) depends, as it does in a milder climate on the harvest of corn, dried fish being the only substitute for bread and fodder among the people of the north. The more simply organized inhabitants of the sea will lead us to make some general remarks.

We have recognized, in the equatorial ocean, a laboratory of nature, where she causes lime-stone to be produced or precipitated by *Molluscæ*, worms, and especially *Polypi*. In the sea which washes the Aleutian islands, animals of the same class are no less numerous, at least, as far as concerns the number of the individuals, and many of the species are not less gigantic than in that zone, but the production of lime-stone does not take place. Among the *Molluscæ*, the cuttle-fish, (*Sepia octopus?*) which is the most remarkable, grows to a size, that really renders it dangerous to the small baydares of the natives, as it is able to overturn them, and justifies, in some degree, the fable of the polypus, which is said to entwine ships in its arms, and draw them to the bottom. There is no great variety of testaceous animals, and if the num-

* Inclined to make comparisons, we observe that Marco Polo mentions, in the 46th chapter of the third book, of the country of Aden (in the torrid zone) that even the horses, oxen, and camels, all eat fish, as no herb appears above the soil, on account of the extreme heat. The cattle rather eat dried than fresh fish.

ber of species is not very considerable, it is made up for by that of the individuals of a few species generally diffused. Several *Balanus*, and the common muscle, (*Mytilus edulis*,) for the most part cover the strand. The muscle, which among us is generally eaten, is here a most dangerous food, which is only taken in case of necessity. It is said to operate, at times, as a most decisive poison; and we were assured that people had often died in consequence of eating it. No *Molluscæ* of these seas can be compared for the production of lime-stone with the *Chama gigas* and other species of the south.

Among the *Zoophytes*, Cuv. the sea-stars (*Asterias*, L.), sea hedge-hog (*Echinus*, L.), and sea-jelly (*Medusa*, L.) are distinguished. The most common sea-star (*Asterias rubens ?*) attains a size of almost a foot in diameter. A *Eurgale* (*Caput medusæ*) is decidedly a different kind from that found at the Cape of Good Hope. The most common sea hedge-hog (*Echinus esculentus ?*) serves for food. The *Medusæ*, and other insignificant animals, supply the whales with sufficient sustenance.* The place of the southern *Lithophytes* is occupied by the *Ceratophytes*; and the north coast of Umnack, in particular, produces several very distinguished species. The fishermen frequently draw up with their lines, from the bot-

* We did not meet with *Clio borealis* in these seas.

tom of the sea, large twigs, six feet long, which, from their near resemblance, they consider to be the beard of a gigantic animal, and which appeared to us to be the skeleton of a sea-pen, (*Pennatula.*)

It now remains for us to describe the people who inhabit the coasts and islands we have noticed.*

It is well known that the Tschukutskoi, settled on the north-east point of Asia, the inhabitants of St. Lawrence Island, those of the opposite coast, and in fact all the inhabitants of the northern coasts of America, beginning from Beering's Straits, on the one side southwards, as far as the Konagi, at Kadjak, and the Tschgatzi, at the back of Cook's inlet, and, on the other side, to the north and eastward, along the Icy Sea, at the mouth of Mackenzie and Copper-mine rivers, as far as the Esquimaux, on the north of Hudson's Bay and Labrador, and to the Greenlanders, and the people, discovered by Ross, in the highest north of Baffin's Bay, belong to one and the same race of men; a race of a decidedly Mongol physiognomy, that of of the Esquimaux, whose Asiatic origin is evident, and whose wandering may easily be followed over

* We observe that we call most of these people and tribes, mentioned by us, by names which they did not give themselves, but which were imposed upon them by strangers. And this is the case with most people on the earth. Thus the word Aleutian seems to be derived from the interrogative particle *allix*, which struck strangers in the language of that people.

the East cape of Asia, and along the coasts of America.

The language, remarkable for its artificial construction, the mode of life, the manners, arts, the very peculiar navigation in leather-boats (Kadjak baydares *), the arms, the costume, are every where the same in the essential parts, and in the drawings given by travellers we can scarcely distinguish the Greenlanders from the Tschukutskoi, or Konägi.

Vater, in Mithridates, vol. iii. part 3. p. 425, hesitates to reckon the inhabitants of the Fox islands, the Aleutians, with G. Forster among the Esquimaux; but they clearly belong to the same race. Dr. Eschscholtz has convinced himself of the essential coincidence of their differing dialects with the original language, and they otherwise entirely resemble their relatives of the same race. This people has evidently wandered from the American continent westwards, to the islands; the most western of the chain have remained unpeopled as those lying in the interior of the Kamtschatka basin.

The language of this branch of people is sufficiently well known to us, chiefly by the schoolbooks in the Greenland dialect, for which we are indebted to the Danish missionaries, and by the Bible, translated into the Greenland and Labradore

* It is remarkable that they are wanting among the northern highlanders of Ross.

languages.* Dr. Eschscholtz has undertaken, with the assistance of an Aleutian who accompanied us, to clear up the Aleutian dialect, and its very intricate grammar. He was resolved to conclude this equally difficult and meritorious work which he had begun; and it is to be hoped, that the necessary assistance of his *protegé* in this labour, will not be withdrawn from him.

In the Aleutian, as well as in the Greenland dialects, there is a remarkable difference in the language of the men and of the women.

The Kamtschadales do not belong to this branch. They are, likewise, a Mongol race, and speak different dialects of a seemingly peculiar language. This people is now almost entirely extinct under its new foreign yoke. (Vide Krusenstern, vol. ii. chap. 8.)

The author is not competent to speak on the Aleutians and the Russian American Company. He would only be able to express his wounded feelings and his commiseration. Whoever, according to established custom, disregards the right of a defenceless people to their native liberty, must allow that, under this rigorous clime, poverty is wretchedness, and the Aleutians are poor and miserable to an unheard-of degree, compared to the more prosperous, robust, and independent

* Mithridates, vol. iii. part 3. p. 432, and Linguarum Index, p. 85.

tribes of the same branch. They are harmless, wretched slaves, who are even now let out without due economy, though not with the same wanton prodigality as formerly, and whose race will soon be extinct. *

Sauer, Davidoff, Langsdorff, Krusenstern, and others, have expressed their opinions on this subject.

We shall indulge in but a few observations on the more northern tribes, the Tschukutskoi, the inhabitants of St. Lawrence Island, and the shores of Kotzebue's Sound; and refer to the Russian accounts, Cook, the narrator of Billing's expedition, Saretscheff and Sauer, and to the account of our voyage. More competent persons have taken it upon them to speak on these tribes.

* Sauer, in the Supplement to his Voyage, gives an extract from the journal of a Russian officer, which speaks of the first Russian hunters on these islands. "They used, not unfrequently, to place the men close together, and try through how many the ball of their rifle-barrelled musket would pass! Gegori Schelikoff has been charged with this act of cruelty, and I have reason to believe it."

In Billing's time, the people of Oonalashka were still distinguished by greater civilization, refinement, and skill in arts, but now no longer.

On the West Indian islands the negro-slaves, not unfrequently, fled into the inaccessible mountains of the interior (*Negres marrons, Cimarrons*). Here, where only the sea affords sustenance, the Aleutians, on several islands, are said to have fled into the mountains.

We have been officially informed, that the number of the Aleutians on the Fox islands was, in the year 1806, 1334 men, and 570 women; and, in 1817, 462 men, and 584 women (?).

We became acquainted with the Tschukutskoi at the same place whère Cook and Billing had been before us. We found their accounts of the manners and customs of this people, as far as we became acquainted with them, very accurate, but must contradict them only in one point; namely, in respect to the advantages ascribed to them above other tribes, in cultivation, strength, a superior stature, and particularly in more European features. We recognized in them only the Esquimaux of the opposite coast, to whom we even thought them inferior in ingenuity. Only some of them might, perhaps, be distinguished by a higher stature.

The Tschukutskoi acknowledge, indeed, the Russian supremacy; but the tribute, which they voluntarily bring to Russian trading places, is only like a duty by which they obtain access to them, and they enjoy the advantages of trade, while their independence remains unimpaired.

As the island of St. Lawrence lies between the two continents, its inhabitants seem to keep a medium between the Tschukutskoi and Americans, but to be more nearly related to the latter. They do not appear to burn their dead as the Tschukutskoi do. We have found skulls on the *plateau* of the island, and in the fragments of rocks at the foot of the montains, but not the monuments made of drift-wood, which are erected on the American coast, to mark the burial place of the dead, on the frozen-ground of the hills, and protect it against wild beasts. It is well known that they wear the ornaments at the corner of the mouth, which distinguish the Esquimaux, from Kotzebue's Sound, to the mouth of Mackenzie River; but they are less general among them, and of a smaller size. They appeared to stand in commercial relations with the Tschukutskoi, and to procure from them, particularly, dresses (*Parkas*) of rein-deer skin, which they use: they do not possess the animal itself. They are rich in morse-teeth and other productions, procured from marine animals, and are fond of trade.

The Tschukutskoi hate the inhabitants of the American coast, with whom they live in enmity and war, and pictured them to us in the blackest colours. In our intercourse with them, we observed only that precaution which becomes a man capable of bearing arms, towards strangers, and which we ourselves used towards them; but never perceived any thing that could authorize in us a suspicion that they intended treachery. Their riches in Russian goods, iron, blue glass beads, &c., struck us very much. If we understood the Tschukutskoi rightly, and may credit them, they procure these goods, like the Tschukutskoi themselves, from Colima. Shall we believe that the trade of the Americans has really opened a way to this market by sea, round the Schelatzkoy-noss, or rather by night, and in the winter season, by means of sledges, over the frequently mentioned isthmus of this promontory?

METEOROLOGY.—MAGNET.

The naturalist of the expedition had only the observation of Troughton's dipping needle; and that only twice, in Chili and St. Lawrence Bay. He can only repeat what is found in Ross's Voyage, Appendix, page 128:—

"We never got any result from this instrument which could be depended on."

APPENDIX

BY

OTHER AUTHORS.

GENERAL OBSERVATIONS.

In the North Sea we saw swarms of *Medusæ* swimming in the sea; for the first time, (the 22d of August, 1815,) we caught a great number of *Medusa capillaris*, L. On the 27th, another large *Medusa* was very frequent, which seemed unknown to us. It was distinguished by the edge of the disk being divided into thirty-two small round incisions, and by the great fluctuation in the form.

The two continued calms, which we had in the Atlantic Ocean, were very advantageous for catching and observing the *Molluscæ*. On the 16th of October we saw for the first time two kinds of *Salpæ;* the one was a *Salpa maxima*, L.; the other a singular kind, composed of two hermaphrodites, externally different, in which I was so fortunate to observe the reciprocal propagation. The one, which I shall call *Hermaphroditus*, has already been described by Forskohl, by the name of *Salpa pinnata*, and is distinguished by its broad elongation, in the front and underpart of the body, which is wanting in the *Hermaphrodita*. The first has a single tolerably long blue stripe on the back and on each side; but the latter, in the same places, five short arched spots in one row. The *Hermaphroditus* does not swim singly about in the sea, in its natural state; but several together, the one, holding itself fast by the points of the elongated excrescence to that of the next, form themselves into a wheel-shaped figure, the elongations forming the radii. In the large channel, through which the water flows that it sucks in, to move the body forwards, it carries a young one, fastened by a navel-string to the inner side of the back: this young one is living, and sucks in and discharges the water at the same time as the one that bears it. It has no elongation on the fore part of the body; and has on each side of the back five blue spots, and is therefore a hermaphrodite.

The *Hermaphrodita*, always single, nourishes in a conical space, in the under part of the body, an innumerable quantity of *Hermaphroditus;* and from eight to ten are born together, holding to each by their elongated excrescence.

During the second calm, on the 24th of October, we caught a number of *Glaucus Atlanticus*, Blumenbach, which seemed to lie quietly on the surface of the water, and on the following day a *Pterotrachea pulmonata*, Forsk., which has a great resemblance with a fish. During all this time we observed the *Gryllus tataricus*, L. lying in great numbers on the water; a propitious wind had happily driven these destructive swarms into the sea. It did not appear to be many days since this had taken place, as most of those which we took up were fresh, and not yet beginning to corrupt. A few years before, they had visited Teneriffe, and occasioned great damage about the town of Laguna.

The day after our arrival at Teneriffe, Chamisso and myself commenced our three days' excursion into the island, early in the morning. Our guide first conducted us into the mountains, on the western point of the island. The rocky walls behind this town were covered with fleshy plants, of which the *Euphorbia canariensis* and *Piscatoria*, *Cacalia Kleinia*, and the *Cactus Opuntia*, which grows entirely wild here, sometimes form entire thickets, being arborescent. Farther on, in a valley, we saw two pretty high dragon trees, (*Dracaena draco*, L.) one of which bore fruit. About two o'clock at noon, we descended into a large vale, and saw the town of Laguna. An aqueduct leads from the mountains into the town. Not a single inn was to be found on this spot; there were four monks' and two nuns' convents in this place. We had incessant torrents of rain all the afternoon; and, though the rain had not ceased on the following morning, we continued our way to Oratava. At first, a broad even path, led us between fertile fields of lupines, through large villages, but it soon became, on account of the frequent clefts in the rocks, only a broad foot-path, consisting entirely of steps. It was also here that the most delightful vineyards were planted, and where we frequently saw the date-tree, (*Phœnix dactylifera*, L.) in gardens fenced in with the American aloe. At noon, we had proceeded so far, that we could behold from an eminence the town of Laguna and its spacious harbour, which was full of ships. But as we had fixed to be on board on the evening of the following day, we did not proceed, but turned back, when it was already getting dark, to seek our night's lodging in a large village. The rain fell in torrents during the night; which washed the streets so clean, that it was very good walking on the following day, and we had already reached Laguna about noon. We took from this place the nearest way along the high road to Santa Cruz, which lasted a few hours. On this road we met two loaded camels, of a dun colour.

Among the sixty-two plants, which we collected during these three days, in such unfavourable weather, we found thirty peculiar to the Canary or Madeira islands, thirty that these islands have in common with Europe, and only two with Africa. The number of the dicotyledones, amounts to forty-six; and that of the monocotyledones to only nine; the first are therefore as five to one. We only found seven kinds of ferns.

On the 14th of November, after we had passed Cape Verd islands, we caught three sharks, (*Squalus glaucus*, L.) which had followed a boat that was sent out to examine a large piece of wood floating on the water. On one of them a small sucking fish was fastened, (*Echeneis remora*, L.) On the 16th, three herons, (as we took them to be,) followed the ship; they appeared to be very fatigued; for one of them tried to seat itself on the back part of the ship, but fell again into the water. On the following day, we shot a duck, which we decided to be the *Anas sirsair*, described by Forskohl, in Egypt. To-day a shark fastened itself again to the hook, which we recognized to be *Squalus carcharias;* two *Echenei* were attached to its body. On the 20th of November, we saw the first large Brazilian Holothuria, (*Physalia*). On the 7th of December, (near the Brazilian coast,) we sailed through a long yellow streak in the sea, several fathoms broad. We drew some of the water up in a pail, and observed that the yellow colour was occasioned by immense numbers of small, thin, yellow stems, about half a line long. Through the microscope we clearly discovered several transverse divisions in each of the round stems. Two days after, we again met with such streaks in the sea; we afterwards saw narrower green stripes, which were formed of myriads of microscopic animals. These were distinguished by no external members, and had a resemblance to a *Planaria.* On the 10th we descried the Brazilian coast; we saw to-day the first frigate birds, (*Pelecanus aquilus*, Lin.)

Though the mountains of the coast of the continent opposite the island of St. Catharine's are not very high, none of us ever succeeded in reaching the top of one of them. The principal cause was the impenetrable wood; we could only ascend as far as the slaves had opened a way with the axe, for fetching wood; if we attempted to go farther, we soon found ourselves raised above the ground, and suspended upon a texture of climbing plants. For in the same manner as the plants of Teneriffe showed a tendency to arborescence, the vegetation in Brazil was remarkably disposed to the rampant and climbing character. Among the two hundred and thirty-seven kinds of plants, which we collected here, were one hundred and twenty-eight *Dicotyledones*, sixty-nine *Monocotyledones*, and forty *Filices*. The proportion of the *Dicolytedones* to the *Monocotyledones* is therefore two to one. The number of the *Filices* is very remarkable, being to that of the *Phenogamous* plants, as one to five.

On the 2d of January, 1816, we saw, to the north of Cape Victoria, on the coast of Chili, great numbers of animals of the dolphin kind swimming in the sea. They were distinguished by having no dorsal fins; and, in swimming, very much resemble sharks, on account of their large pectoral fins. They have a pretty long snout; the upper half of their body is brown, and the under half white. They did not swim as fast as dolphins, moved the whole body, and also came up to the surface of the water to draw breath.

We collected three hundred and eighteen kinds of the autumnal plants, in Chili, among which, one hundred and sixty-eight *Dicotyledones*, thirty-five *Monocotyledones*, and only fifteen *Filices;* consequently, the proportion of the *Dicotyledones* to the *Monocotyledones* is five to one. From the port of Talcaguano to the town of Conception we could observe three distinct Floras or kinds of vegetation: the first on low hills, covered with clay, towards the sea, was distinguished by its beautiful myrtle groves, and thickets of *Guevina*, in the shade of which grew large parasite plants, *Lianes*, for example, *Lapageria*, *Lardizabala*, &c. The more unfruitful spots of these eminences, were made green by the shrub-like *Pitcairnia coarctata*, with long prickly leaves. The second Flora was in the large sandy valley, scarcely rising above the level of the sea, and extending from Talcaguano to Conception; *Œnotheræ* and *Corymbiferæ* were chiefly in blossom at this season. The third Flora commenced at Conception, on the mountains of mouldering granite. We hardly touched upon this at all.

In Easter Island, we could not find, all along the west coast, any of the numerous statues, described by all navigators who have visited this island; only a pedestal seemed to remain; we saw four upright black blocks on the S. E. coast, but at a great distance, which we must consider as such statues. Of the pierced and extended ear-laps, which were formerly in fashion, we only saw one instance, in an old islander. The stone on the part of the shore where we landed was old brown lava. The tattooing of the face consisted chiefly of one stripe, running all round, near which were several round spots, at equal distances from each other. Some had the face entirely tattooed, so that only a few red spots remained, which showed the colour of the skin. The lips of all we saw were dyed either blue or black. Many of the people were thickly marked with black dots, from the instep to the knee. Of the domestic animals, we only saw one fowl. On the 7th of April we caught a *Salpa bipartita*, Labill. which was very different from all other *Salpæ*. At the same time I likewise obtained the only true marine insect hitherto known, which has much resemblance to the *Velia*, Latr. It runs with great rapidity on the surface of the sea (like the *Hydrometra rivulorum*, F.) and lives in all seas under the torrid zone.

The inhabitants of the Penrhyn islands were not tattooed; some of them had their hair cropped, and others wore a wreath of black feathers round their head; some of them also wore a small mat on the shoulder, to serve as a mantle. In each of the boats, which are furnished with an out-rigger, was an elderly man, who had a whole cocoa-leaf hung round his neck, some of the foliola, on one side, being fastened together. On the out-rigger, they had tied a number of lances of cocoa-wood, about two fathoms long, which they first declined to barter for iron. They kept up an incessant loud noise, distorted their eyes, and made other savage motions, to make themselves appear terrible. Their boats were very simple, resembling a trough, and composed of many pieces, but did not seem adapted to sailing; in general, there were six men in each. They seemed to call iron (as on Easter Island) *Hoïo:* they appeared not to understand the word taboo, or rather did not know what we meant by using it, as the word is, to them a very serious one. *Waihini* seemed, in their language, also to be a denomination for a woman. They brought us nothing eatable, with the exception of green cocoa-nuts; they, however, seemed to be very well fed.

On the 13th of May, (after we had been only two days in the northern hemisphere,) we harpooned a dolphin for the first time, (*Delphinus Delphis*, Lin.) The head ended in a long snout; on each side of its jaw we counted twenty-three teeth, that is in all ninety-two. This day we likewise, for the first time, caught a *Velella*, the internal gristly shell of which, went from the front angle of the left side, in the under membrane, to the back angle of the right side, (placing the animal with one of the longer sides of the under membrane towards the spectator.) This *Velella* was about three inches long, and rather narrow; the shell was of a light brown colour, the membraneous edge of the sail scarcely blue. We caught, at the same time, a *Porpita*. On the 1st of June, during a calm, we caught a second *Velella*, the shell of which and the sail in the membrane, had an opposite direction to the first, which we caught on the 13th of May. The largest specimens were only an inch and a half long; the feelers, which in the former were light brown, and blue only at the end, had here a blue colour from the basis, and the points reddish yellow. The shell was of a very light yellow colour, and only the lower membrane tinged blue on the edge. On this day, we also caught a *Glaucus*, which differed from Blumenbach's *G. Atlanticus* only in being smaller. The following day we saw a grey gull, (*Procellaria furcata*, Pall.) flying about in great numbers.

ON THE CORAL ISLANDS.

I. THEIR ORIGIN.

THE low islands of the South Sea and Indian Ocean mostly owe their origin to the operations of several species of coral. Their situation with respect to each other, as they often form rows, their union in several places in large groups, and their total absence in other parts of the same seas, make us conclude, that the corals have founded their buildings on shoals in the sea; or, to speak more correctly, on the tops of mountains lying under water. On the one side as they increase, they continue to approach the surface of the sea, on the other side they enlarge the extent of their work. The larger species of corals, which form blocks measuring several fathoms in thickness, seem to prefer the more violent surf on the external edge of the reef; this, and the obstacles opposed to the continuation of their life, in the middle of a broad reef, by the amassing of the shells abandoned by the animals and fragments of corals, are probably the reason that the outer edge of the reef first approaches the surface. As soon as it has reached such a height, that it remains almost dry at low water, at the time of ebb, the corals leave off building higher; sea shells, fragments of coral, sea-hedgehog shells, and their broken off prickles are united by the burning sun, through the medium of the cementing calcareous sand, which has arisen from the pulverisation of the above mentioned shells, into one whole, or solid stone, which, strengthened by the continual throwing up of new materials, gradually increases in thickness, till it at last becomes so high, that it is covered only during some seasons of the year by the high tides. The heat of the sun so penetrates the mass of stone when it is dry, that it splits in many places, and breaks off in flakes. These flakes, so separated, are raised one upon another by the waves at the time of high water. The always active surf throws blocks of coral (frequently of a fathom in length, and three or four feet thick) and shells of marine animals between and upon the foundation stones; after this the calcareous sand lies undisturbed, and offers to the seeds of trees and plants cast upon it by the waves, a soil upon which they rapidly grow to overshadow its dazzling white surface. Entire trunks of trees, which are carried by the rivers from other countries and islands, find here, at length, a resting place, after their long wanderings: with these, come some small animals, such as lizards and insects, as the first inhabitants. Even before the trees form a wood, the real sea-birds nestle here; strayed land-birds take refuge in the bushes; and at a much later period, when the work has been long since completed, man also appears; builds his hut on the fruitful soil formed by the corruption of the leaves of the trees, and calls himself lord and proprietor of this new creation.

II. FARTHER FORMATION AND PECULIARITIES OF THE CORAL ISLANDS.

In the preceding account we have seen how the exterior edge of a sub-marine coral edifice first approaches the surface of the water, and how this reef gradually assumes the properties of land; the island, therefore, necessarily has a circular form, and, in the middle of it, an enclosed lake. This lake, is, however, not entirely enclosed; (and it could not be, for, without supply from the sea, it would soon be dried up by the rays of the sun,) but the exterior wall consists of a great number of smaller islands, which are separated from each other by sometimes larger, sometimes smaller spaces. The number of these islets amounts in the larger coral islands to sixty; and between them it is not so deep, but that it becomes dry at the time of ebb. The interior sea has, in the middle, generally a depth of from thirty to five and thirty fathoms; but on all sides, towards the land, the depth gradually decreases. In those seas, where the constant monsoons prevail, where, consequently the waves beat only on one side of the reef or island, it is natural that this side of the reef, exposed to the unremitting fury of the ocean, should be formed chiefly by broken off blocks of coral, and fragments of shells, and first rise above the element that created it. It is only these islands, respecting the formation and nature of which we hitherto know any thing with certainty; we are still almost entirely without any observations on those in the Indian and Chinese sea, which lie in the regions of the six months' monsoons. From the charts given of them, it is to be inferred that every side is equally advanced in formation. The lee-side of such a coral reef in the Pacific Ocean, which is governed by the constant monsoons, frequently does not show itself above the water, when the opposite side, since time immemorial, has attained perfection in the atmospheric region; the former reef is even interrupted in many places by intervals tolerably broad, and of the same depth as the inner sea, which have been left by nature like open gates for the exploring mariner to enter the internal calm and secure harbour. In their external form, the Coral islands do not resemble each other; but this, and the extent of each, probably depends on the size of the sub-marine mountain tops, on which their basis is founded. Those islands, which have more length than breadth, and are opposed in their greatest extent to the wind and waves, are richer in fruitful islands than those whose situation is not so adapted to a quick formation. In the large island-chains there are always some single islets, which have the appearance of high land: these lie upon an angle projecting into the sea, are exposed to the surf from two sides, consist, therefore, almost entirely of large blocks of coral, and are destitute of smaller fragments of shells and coral sand to fill up the intervals. They are, therefore, not adapted to support plants requiring a depth of soil, and only afford a basis to high trees, provided with fibrous roots, (as the *Pisonia*, *Cordia Sebastiana*, L. *Morinda citrifolia*, L. and *Pandanus odoratissimus*, L.) which, at a distance, give to these, always very small, islands the form of a hill. The inner shores of the island, exposed to the surf, consist of fine sand, which is washed up by the tide. Between the small islands, under their protection, and even in the middle of the inner sea, are found smaller species of coral, which seek a quiet abode, form in time, though very slowly, banks, till they at last reach the surface of the water, gradually increase in extent, unite with the islands that surround them, and at length fill up the inner seas; so that what was at first a ring of islands, becomes one connected land. The islands which are so far formed, retain in the middle a flat place, which is always lower than the wall that surrounds them on the banks; for which reason pools of water are formed in them after a continued rain—the only springs and wells they possess. Among the peculiarities of these islands is, that no dew falls in the evening, that they cause no tempests, and do not check the course of the wind. The very low situation of the country sometimes exposes the inhabitants to great danger, and threatens their lives, when the waves roll over their islands, if it happens that the equinox and full moon fall on the same day, (consequently when the water has reached its greatest height,) and a storm agitates the sea at the same time. These islands are said to be also shaken by earthquakes.

ON THE

NATURE OF THE ROCKS

ON THE COAST OF

NEW CALIFORNIA, THE ISLAND OF OONALASHKA, AND THE COASTS OF BEERING'S STRAITS,

ACCORDING TO THE OBSERVATIONS MADE THERE, AND MINERALS COLLECTED BY DR. ESCHSCHOLTZ.

BY

MORITZ VON ENGELHARDT.

Dr. Eschscholtz gave me the minerals which he had collected, that I might examine them, and had the politeness to give me information respecting the situation in which each specimen was found; this was the origin of the following remarks, and the list accompanying them.

M. V. E.

NEW CALIFORNIA.

The northern tongue of land running into the sea, on which the little fortress of St. John, at the entrance of the bay of San Francisco lies, consists of serpentine rock, which is also found farther south on the steep shores of the sea coast. With it are mixed, amianthus, scaly talc, magnetic pyrites, and bronzite, (or Schiller Spar), minerals which in other countries usually accompany the serpentine in a similar manner, and afford here an additional proof of the regular course of the process, of the formation of the earth. The same is shown by the manner of the stratification. It is well known that in most primitive rocks, the serpentine appears on the outside, that is, there where its last layers, (that uniformly cover the other rocky strata,) border on the floetz mountain; in Saxony for instance, at the circumference of the white stone, the slaty covering of which adjoins rocks of later formation; on the south side of the Swiss Alps, (near Aviglia and Yvrea), where the plain of Piemont begins with compound mountains (Schütthügel); in Silesia, in the Zobten mountain, which extends far into the level country, and in the same manner on the coast of New California. Not only does a piece of serpentine that divides into flakes, which was thrown upon a sand-bank by the current coming from the interior of the bay, prove that it laid between slates, because the serpentine could not change its otherwise compact mass, except by this means; but the clay-slate itself appears in the low rock at the foot of the serpentine wall, and these border on sand-stone, and conglomerate, the cliffs of which occupy the narrow border of the coasts. The vicinity of the sea is common to the serpentine of New California, with the serpentine at Cape Lizard, on the Shetland islands, and on the Cordillera of the coasts of South America. If we consider that the sea formerly covered those countries which contain the floetz mountains, which are rich in petrefactions, and admit the hypothesis, that primitive mountains, which were not covered by the floetz mountains, rose as islands above these seas, it appears how the serpentine of what is now the interior country formerly lay equally situated on the coasts; an analogy, which, if thoroughly investigated, may lead in future to important conclusions respecting these and other kinds of rock, in the history of the formation of the surface of the earth.

OONALASHKA.

Dr. Langsdorff reports, in his voyage round the world, that this island consists of granite and porphyry, as Kamtschatka, and the whole chain of the Aleutian islands, from the north-west coast of America, contains nothing but primitive rock. From the beautiful series of specimens of rock which have been brought from Oonalashka, and in which only the wood-stone (*Lythoxylon*) is wanting, which Dr. Eschscholtz, however, frequently saw there, it appears that older sand formation prevails with amygdaloid, porphyry, claystone, and jasper. Lavender-blue and brown-red clay-iron-stone forms the paste of these kinds of rocks, which change into each other. The amygdaloid contains calcspar, much green earth, stilbite, glassy felspar, also small nests of red iron-stone; the porphyry, properly only a condensed amygdaloid paste, sometimes becomes jasper, and has, besides the minerals already mentioned, also small crystals of common felspar. Where green earth was accumulated, the colour of the stone changed into grey green; where flinty earth and iron prevails, it increases in hardness, where they are wanting it becomes clay, and changes into sandstone, resembling that of coal.

These kinds of rock (most of which have a striking similarity with those which are met with in the interior of the same formation, upon the Nahe, on the left bank of the Rhine, and in the north of Germany) were found partly on the east, partly on the west side of Captain's Harbour, (a bay of the north coast,) in steep pointed rocks, which are subject to continual changes. Where travellers formerly saw and sketched conical summits, (as Saritscheff did,) there were now saddle-shaped hollows; the fragments of the former summit covered the sides; where Dr. Eschscholtz, during his first stay at Oonalashka, in the year 1816, had seen these hollows flat, he found them farther hollowed out on his second visit to Oonalashka, in 1817, and the formerly low round tops of the lateral boundary changed into small peaks. Earthquakes had not now been the cause of those changes, but, probably, the unequal solidity of the stone composing the masses of rock which are piled one over the other. In places where I had opportunities of observing these formations, I saw, analagous to the change of the clayey, flinty, and sandy places, in one and the same layer, masses of crumbly clay, slate, loose sandstone, and conglomerate, firm porphyry and amygdaloid, alternate with each other; and of the two last, grotesque rocks, and rugged walls, produced by the washing away and sinking together of their softer intervenient layers, rising among dreary ruins.

Not less interesting than the kinds of rocks just mentioned is the appearance of the porphyry-slate and rocks resembling basalt, at Oonalashka. Do they belong here to the formation of the older floetz sandstone, or to the floetz trap? These questions, which force themselves upon us, cannot indeed be satisfactorily answered for want of knowledge of the stratification of the two kinds of rock, but the supposition may be hazarded, that these rocks belong to the floetz trap. Indissolubility, greater hardness, by which the piece No. 30. is distinguished from real basalt, seem to be owing to quartz and augite, which here form a chief ingredient of the rock, though otherwise only sprinkled, like porphyry, in basalt. But that this stone contains augite, speaks for the affinity with the real floetz trap, which is farther confirmed by the porphyry slate, which (occurring only in this formation) likewise contains this fossil.

If the supposition which adopts at Oonalashka members of the family of floetz trap, is correct, it will be a confirmation of the near relation that has been often noticed elsewhere between these formations and volcanoes; an affinity which must be equally interesting to the Neptunists and to the Vulcanists, though there is but little hope that their contest will be decided on this remote field.

From the shores of the harbour of St. Peter and St. Paul, in Kamtschatka, green, yellow, and brown-red jasper was brought, which is stated to lie there in horizontal layers; we also have from the peninsula, without any knowledge of the place where it was found, a beautiful calc-spar crystallization, (H.'s equï-axe) on scales of chalcedony, and brown semi-opal, the fragment of a kernel of the amygdaloid. The southern Kamtschatka, therefore, probably contains the same kind of rocks as Oonalashka; and as, according to Mr. Langsdorff, the Aleutian islands, which lie between both, resemble Oonalashka in the exterior as in the volcanoes, it is to be conjectured that the whole chain consists of floetz rock.

BEERING'S STRAIT.

Saline marble, resembling that which forms the ruined rocks of St. Lawrence Bay, is regularly found as strata in mica slate, and this probably is likewise the case on the Tschukutskoi coast, as the specimens brought from thence contain white-grained calcareous stone sprinkled with silver-white mica, and graphite is also found in the neighbourhood, which generally belongs to the mica slate. Of the kind of rock itself, we have a specimen from Kotzebue's Sound, lying opposite, where it may be found in the neighbouring mountains. The mica is silver-white, like that which the limestone contains; may not, therefore, the hypothesis be allowed, that the primitive rock continues from Asia to America, and that both continents were once united at Beering's Strait? As the one coast (the Asiatic) is said to be steep, the opposite one flat, they are to each as the banks of a river and side valleys, forming by running waters. The supposition of a later separation is therefore not contradicted by the nature of the strait.

No direct observation has informed us what kinds of rock occupy the space between Beering's Strait and the Aleutian islands; but as on the north side of Oonalashka a layer of gneiss-syenite was found, and in Kotzebue's Sound a very remarkable porphyritic syenite, these formations may perhaps serve as a basis for the floetz mountains of the Aleutians. In this case the extended basin between the above mentioned island-chain and New California, appears to be an inlet bounded by two groups of primitive rock, and filled with floetz rock, whose similar formations (sand-stone, conglomerate, jasper) rise on both sides of the basin. Their stratification seems to be regular; for if you draw a line between the conglomerate rocks and the serpentine near Saint John, to the north side of the Bay of San Francisco, where Dr. Eschscholtz saw rocks of a red-brown colour, and which were considered to be conglomerate; and extend the line to the S. W., it meets the volcanic Sandwich islands, and has exactly a similar direction with the Aleutian islands, from Alashka to Atha.

It is remarkable, that as in many other places where masses of land are separated, here too, volcanic islands lie before the entrance of Beering's Strait. Has the whole chain of the Aleutians risen from the sea like the island that rose in the year 1795 or 1797, near Umnack? Or are here only summits of a chain of mountains, the bases of which are at the bottom of the ocean? Or are they remains of a rocky dam that has been rent asunder? The answer to these questions might perhaps be obtained, if the nature of the rocks of the two coasts of Asia and America, from Beering's Straits to the chain of the Aleutians, and this chain itself were examined; but of this we can have little hopes, so long as, in voyages of discovery, even on accessible coasts, attention shall be paid to the collection of plants and animals, but no regard had to the formation of the earth, though a firm basis of physical geography is to be sought in this alone. Till the importance of geognosis is more generally acknowledged, and accurate observations made in remote parts of the earth, we can form only conjectures on the nature of it, drawn from a comparison of circumstances of its structure that are merely hinted at, with others more accurately determined.

LIST OF THE MINERALS

COLLECTED BY DR. ESCHSCHOLTZ,

AND PRESENTED TO THE CABINET OF NATURAL CURIOSITIES OF THE UNIVERSITY OF DORPAT.

FROM THE COAST OF NEW CALIFORNIA.

1. *Common Serpentine* crossed in all directions by blueish and greenish-white steatite, which encloses veins of amianthus, (*Asbestus?*): contains tombac-brown bronzite in single folia: slightly affects the magnetic-needle. — From the rock of the sea coast, south of the fortress of St. John, at the entrance of the Bay of San Francisco.

2. *Common Serpentine,* intermixed with scaly talc: contains magnetic pyrites, interspersed in small crystals, also small folia of bronzite. — Is found with No. 1.

3. *Serpentine* approaching the noble, with greenish-white earthy talc in undulating lamellæ. — From a sand-bank in the Bay of San Francisco.

4. *Serpentine* which approaches the noble. Fissures separate the stone into angular and roundish pieces, the superficies of which is spread over with a thin layer of chalcedony, which is covered with acicular arragonite. The amianth veins which

traverse the serpentine are not heaved by the fissures which cross them. — From the sand-bank.

5. *Scaly-talc*, leek and olive-green; of a greasy aspect, tolerably soft, very friable, adheres strongly to the tongue. — From the rocks of the coast.

6. *Scaly-talc* and *Serpentine*, lying in flakes over one another, crossed by amianth. The mountain-green scaly-talc, clearly an intimate mixture of amianth, (*Asbestus?*) and steatite. — From the rocks of the coast.

7. Black contorted *Clay-slate*. The surfaces of the lamellæ, when newly broken, are shining, striped brownish-grey. — Low rocks along the sea-coast, at the foot of the serpentine-stone wall.

8. Greenish-grey *Sand-stone* and *Conglomerate*. Quartz grains united without visible cement, form the chief mass, in which fragments of clay-slate and heliotrop are imbedded.

9. *Heliotrop*, a large fragment from the conglomerate, which is still united to it.

10. Compact *Green-stone*, with sprinkled iron pyrites. — Fragment from the conglomerate.

11. Brown yellow *Sand-stone*, composed of quartz grains and many fragments of clay-slate. — Forms rocks which run into the sea at the foot of the serpentine rock.

12. Ferruginous, reddish-brown common *Jasper*, having a metallic lustre, and many small fissures, which are filled up by calcareous spar. — From the sand-bank — the bay.

13. *Jasper*, passing into brown iron-stone. The surface, where broken, of a metallic lustre. — With the conglomerate, No. 8., rocks which project out of the sea.

14. *Conchoidal Horn-stone*, flamed, flesh-coloured, greenish-grey, and greyish-white. Finely sprinkled with iron-pyrites. — Found loose near No. 13.

FROM OONALASHKA.

15. Greenish-grey small-grained *Sand-stone*, resembling coaly sand-stone (*Steinkohlen sand-stein*): with fragments of clay-slate and pieces of a fossil, too small to be distinguished, which is dark-olive, and leek-green, transparent, soft, and having a conchoidal fracture. — Thrown up by the sea, near the settlement of Illuluk, in Captain's Harbour.

16. Reddish and greenish-grey *Sand-stone*. Finer grained and firmer than the preceding. Reddish-grey calcareous clay iron-stone (*Eisenthon*), is the cement of the quartz grains. — From the shores of Captain's Harbour.

17. Dark lavender-blue *Clay Iron-stone*, with a fine-grained uneven fracture, a tendency to slatiness, and the surface ochreous; lying on the mountain east of the settlement.

18. *Clay Iron-stone* and *Green Earth* mixed, so that in some places the one predominates, and sometimes the other. — Situated on the west side of Captain's Harbour.

19. *Amygdaloid;* mass of clay iron-stone, like No. 17. only firmer. It has longish pores, of which some are empty, others filled with green earth and lithomarge. Glassy felspar in small crystals. Globular masses covered by pieces of green earth, passing into apple-green lithomarge. — Over No. 17.

20. *Amygdaloid;* as the preceding, but without felspar, also with nests of compact red iron ore, enveloped in the green earth. — As No. 19.

21. *Amygdaloid;* the green earth finely sprinkled; the calcareous spar in small masses and nests, rare; glassy felspar is abundant; hence the whole is like porphyry. — From the shores of Captain's Harbour, near Illuluk.

22. *Clay Iron-stone,* separating into spherical masses, almost quite compact, spotted by green earth; rendered like porphyry, by crystals of felspar. — On the shore near the settlement.

23. *Porphyry.* Blueish, brown, red mass of clay iron-stone, with felspar crystal. — As No. 22.

24. *Porphyry.* Dark red brown, compact and quartzy paste, with brown-red jasper veins, greenish felspar crystals, and single small masses of zeolite.

25. Brown-red *Porphyry,* with small crystals of compact felspar, tinged by green earth. — As No. 22.

26. *Porphyry-slate.* Greenish-grey, with many white glassy felspar crystals, raven black and

dark olive-green augite crystals and grains, enclosing iron pyrites. — Layers from Captain's Harbour.

27. *Porphyry*, weather-worn; as are also the felspar crystals, which are earthy and greenish-grey. — On the western side of Captain's Harbour, with No. 18.

28. Grey-green *Porphyry*, full of spots of green earth and felspar crystals. — Under No. 27.

29. *Amygdaloid*. The mass by exposure is become earthy and greenish-grey, contains calcareous spar, covered by stilbite. — Layers from Illuluk.

30. Stones resembling basalt. Brownish black, fine-grained, tolerably hard, lustre glistening, sprinkled in with augite, and porphyrylike in crystals, (*Hauy's Pyroxène hêmitrope.*) Specific gravity 2,701. — From the foot of the so called great volcano on the west coast of the island.

31. Compact liver-brown and bluish-black spotted stone, with fine pointed crystals of a brown-red, very soft, undefinable mineral, and with a finely-sprinkled felspar, in incipient decomposition. Fracture, straight lamellæ. Blueish earthy mould covers the surface of the clefts, and red hematite penetrates into the mass from without. Specific gravity 2,621. — Layers from the west side of Captain's Harbour.

32. Yellowish white *Clay-stone*, with red jasper

veins, which contain small drusy cavities of Mesotype. — Layers from Illuluk.

33. Brick-coloured common *Jasper*, allied to eisenkiesel. — As No. 32.

34. Small *Rock Crystals.* — From cavities of the almond-stone, which lies east of Illuluk.

35. Blueish-white almost compact quartzy *Felspar Rock*, with nests of reddish-white porcelain earth, and much, finely-sprinkled iron pyrites. — Lies in beds at Illuluk.

36. *Gneiss Syenite.* — Layers from the northern sea coast, near Captain's Harbour.

FROM ST. LAWRENCE BAY.

37. White small-grained *Calcareous Spar*, with silver-white scaly mica. — Forms rocks on the south end of the bay.

38. *Calcareous Spar*, sprinkled with plumbago. A flat mass, which appears to have filled up a cleft.

40. and 41. *Foliated Plumbago.*

FROM KOTZEBUE'S SOUND.

42. Silvery-white *Mica-slate*, with noble garnet in small crystals.

43. *Porphyritic Syenite*, red mass of felspar, with tolerably large crystals of hornblende and felspar. — Both are from the beds of the Rocky Island in the Sound.

FROM KAMTSCHATKA.

44. Greenish-yellow *Jasper*, resembling hornstone.

45. Red *Jasper*. — From the shore of the harbour of St. Peter and St. Paul, where the red jasper is imbedded in red, and with this is stratified horizontally.

46. Crystallized *Calcareous Spar*, (H.'s inverse), on grey-blue chalcedony and brown-semi-opal. Fragment of a large kernel of the amygdaloid. — Locality unknown.

FROM THE ISLAND OF OWHYEE.

47. *Scoriaceous Lava*, resembling the dross of a forge. Near the under surface of a longish quadrangular specimen, a more compact layer of lava runs through the porous mass, and which in one half of the piece repulses the north pole of the magnetic needle, and attracts it in the other.

48. *Scoriaceous Lava*, with a very shining surface, almost compact. Strikes fire with steel, and has no influence at all on the magnet. — Both pieces from a stream of lava from the Mouna Wororai, on the west side of this island.

FROM THE ISLAND OF GUAHON.

49. Compact *Lime-stone*, flamed, flesh-coloured, and reddish-white. Has a fine texture almost even in the fracture, and several faint traces of petrifaction. — From steep rocks on the S. W. coast.

Fig. 1.

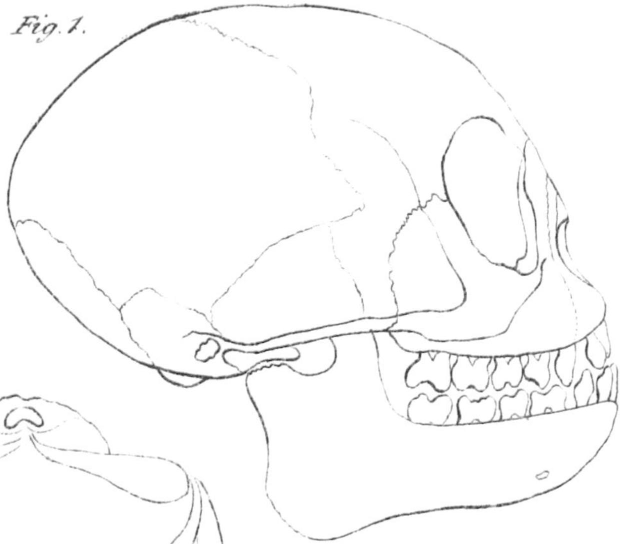

Fig. 2.

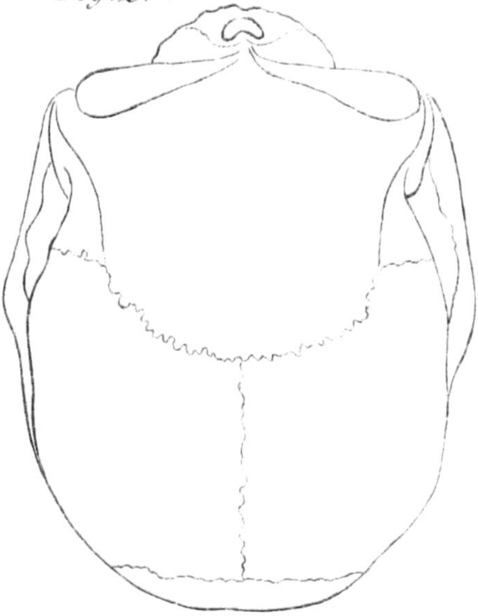

Fig. 3

Fig. 1. Side View of the Skull of the Presbytis mitrata:
Fig. 2. The same from above:
Fig. 3. The Palm of the left fore foot (or hand) of the same.

DESCRIPTION

OF

A NEW SPECIES OF MONKEY.

(PRESBYTIS MITRATA.)

BY FREDERICK ESCHSCHOLTZ.

(WITH A DRAWING.)

THE generic character is: *Angulus facialis* 60 *graduum. Sacculi buccales nulli. Nates tyliis instructæ. Cauda elongata. Antipedes genu attingentes.*

This monkey was brought us for sale, by the inhabitants of Sumatra, on our passage through the Straits of Sunda; it was tame, but lived scarcely a day on board. As, on classing the monkeys, an essential difference was found to distinguish it from all the species hitherto known, I thought proper to assign to it a peculiar species under the name of *Presbytis,* on account of its resemblance to an old woman, with a cap on her head. Its length, from the head to the beginning of the tail, is about a foot and a half. The back is covered with finely-curled (or wavy) hair, two inches long; which is of a yellowish-white colour, at the roots, and towards the tip, a blueish-grey; the back part of the head is also covered with a similar long hair, of the same colour, which stands erect, and gives the monkey the appearance of having on a fur cap. This greyish hair is bordered by a black arched band, formed of hair three quarters of an inch long, and extending from the place where the zygoma joins the coronal bone, to the upper edge of the ear, and from thence directly across the head. The whole space between this band and the eye-brows is thickly covered with yellowish hairs, of which those nearest the band measure a full inch; but those near the eye-brows only half an inch. Only one very raised place in the middle of the forehead, lying above the eye-brows, is almost bare of hair, and shows the black skin. A few single straight black hairs, nearly an inch long, may be considered as the eye-brows, which, however, extend over the nose. The *concha* of the ear has not a very broad edge; its skin is of a reddish colour, and it has on it long yellowish hair; an ear-lap is also perceptible. The colour of the skin of the face is almost black; both eye-lids however, (the upper is particularly broad,) are reddish. The lips are covered with short whitish hair.

The hair on the under side of the body measures almost two inches, and is white. The tail is longer than the body, covered with pretty long hair every where, at the top blueish grey, beneath greyish yellow; at the end is a tuft of yellow hair an inch and a half long. The arms of the animal reach almost to the knee when standing erect; the hair on the upper side is whitish grey, on the under side of the upper arm white and woolly; on the fore-arm yellowish, straight and close. The hands are covered on the back with reddish-brown hair, (as far as the last joint of the finger,) the fore-fingers are particularly long and narrow; but the thumb is very short, as it reaches only to the half of the middle of the hand. The two middle fingers are much larger than the two others. The nails of the fore-fingers are long semi-cylindrical, a little bent, which makes them have a resemblance to claws, but round, and in the form of a nail at the tip. Only the nail of the fore-finger is obliquely blunted, which gives it a small point on the inner side. The nails of the thumb are short, raised, depressed and rounded off at the edge. The skin of the bare palm of the hand is reddish. The long white hairs of the hind feet have a woolly appearance; the hands on the back are here also covered with reddish-brown hair; they are longer than the fore-hands, with which they agree in every other particular.

On both sides of the rump are two roundish bare callous places, of a yellowish-brown colour.

With respect to the form of the face, it has only to be observed, that the zygoma projects very much, while the nose is very little, and scarcely raised; seen in profile, the projection of the forehead, the nose, the upper jaw, and the gums of the under jaw, make almost a straight line.

I can add a few words respecting the skeleton: the skull is pretty round, the jaw-bone not much projecting, as the facial angle is sixty degrees; on the sagittal suture, no trace of a crista; the nasal-bone very small, triangular, two lines long; the inter-maxillary bones, begin already above the nasal bones; there are only twenty-eight teeth in my female specimen, which, to judge from the blackness of them, was already full-grown; the corner teeth scarcely larger than the others. There are twelve pretty short vertebræ of the back; the vertebræ of the loins, are seven in number, long, and with broad transverse processes; the os sacrum consists of three vertebræ, of which only two join the os ilium, and, lastly, the tail is formed of twenty-eight longish joints, without processes. Of the twelve ribs, seven are primary, and five secondary; the two lowest secondary ribs are not attached to the place of junction of two vertebræ, but are fixed to the body of one. Of the shoulder-blade, it is also to be observed, that the coracoidal process, which is here long, is divided as a separate bone; but is rendered only half moveable, by a narrow gristle. The sternum consists of five bones, lying longitudinally.

REMARKS

ON THE

PHYSIOLOGY AND NATURAL HISTORY

OF THE

MEDUSÆ, VELELLÆ, AND PORPITÆ.

BY

FREDERICK ESCHSCHOLTZ.

THOUGH the number of authors who have written on these animals is not small, yet the more accurate observer often finds many things which he may add to what is already known, particularly in what relates to their mode of life. These three kinds of animals clearly belong to one order: 1. Because they remain nearly passive, swimming on the surface of the sea in search of prey; 2. Because, besides the body adapted for swimming on the surface, they have as chief organs, the innumerable stomachs and mouths; and, 3. Are all furnished with particular arms for seizing their prey. They live in the warmer regions of all seas; and do not appear to go beyond the fortieth degree of latitude.

I. MEDUSÆ, PHYSALIA.

As we have hitherto been in error and doubt respecting many parts of these animals, and as the mode in which they procure food will doubtless interest many persons, I will here give a connected account of them.

The oval bladder, which forms the body of the animal, and keeps it swimming on the surface of the water, by the air which it contains, may be so contracted on the back by means of muscles, as to form a sort of comb. By the assistance of this comb the medusa catches the wind, and is driven about on the surface of the water. The long spiral arms, which are furnished on one side, from top to bottom, with kidney-shaped suckers, hang down, unrolled behind, and follow the floating bladder; as soon as a fish, or any other animal, touches them, it is caught by the suckers, benumbed by the corrosive fluid, and brought near to the bladder by a spiral contraction of the arm, in which a large entirely muscular feeler (*tentaculum*) growing to the basis of the arm, but free at the point, seems to cooperate. When the prey is thus brought near the bladder, innumerable tubes, provided with suckers, (real stomachs,) await it, seize it on all sides, and suck in the soft and soluble parts. Thus satiated, these tube-shaped stomachs swell and contract in an extraordinary manner, the contents are seen to shine through, like dark grains; while quietly digesting these they pay no regard to any new prey that may approach, but leave it undisturbed to their hungry neighbours.

The long arms seem to grow out singly, as one may always be distinguished as the thickest and longest. In some very large individuals of the *Physalia glauca* we observed, indeed, two remarkably large arms, yet one was always larger than the other. They are all provided at the root with the above-mentioned muscular feeler. The tubercles described by authors on the proboscis of the bladder of the *Physalia glauca*, which is found in abundance near the Cape of Good Hope, are nothing more than little stomachs, that have not attained their complete growth. I have convinced myself by the examination of some full-grown specimens. In one of them, which must have made a good booty just before, not only all the stomachs hanging to the middle part of the bladder, but also the above-mentioned tubercles on the proboscis were filled with a reddish grainy mass; besides, they had all, like those stomachs, yellow points, and those nearest to the bladder of the body, could not be taken for any thing but real stomachs, on account of their whole form and their yellow funnel-shaped mouths. Besides these organs, there are three round bundles of little pale threads hanging down; on nearer examination it was found that each bundle consisted of two kinds of threads, namely, of longer cylindrical pointed threads, and of shorter ones furnished with a sucker. As a longer thread of a blueish colour always hung (in the *Physalia glauca*) close to the shorter one, I imagine that these two threads represented an unformed young one, hanging to the mother, and that the longer thread was the first arm for seizing prey, the shorter one a stomach; it is not necessary for the bladder to get filled with air, as the young one is carried by the mother.

Wishing to see whether the small cavity at the back process of the *Physalia glauca* was really an opening closed by a muscular valve, I took the bladder in the middle, pressed the air towards the end of the back process, and saw how a small round opening was formed in the place of the cavity out of which the air pressed. When I left off pressing, the opening closed again; the bladder became quite slack by the air being pressed out.

II. VELELLÆ.

The flat body of the velellæ, swimming on the surface of the water, has the appearance of a parallelogram with rounded corners; its exterior skin is very soft, and incloses two transparent gristles, joined together in the middle, and forming an ellipse, with concentric stripes. These are a little convex, and lie in the body in a diagonal line. On them stands a transparent gristly semicircular sail, also in a diagonal direction, and in such a manner, that its ends are turned to the exterior edge of the body; at the top, in the middle, it runs out in a small point, and is entirely surrounded by a narrow (probably muscular) skin, in such a manner, however, that it does not cover the point of the sail; beneath, this skin does not attach itself to the body, but remains free. On the under-superficies of the body, there is in the middle a large stomach, in the form of a bottle, which is surrounded by innumerable smaller stomachs, as far as the elliptic gristly mass extends. The whole border of this gristle is covered with pretty long thin *tentacula*, which, where the gristle lies near to the edge of the body, project from under it.

The large middle stomach (which is always considered as the only one, by authors) seems to swallow very small animals, as we sometimes found it filled with remnants of them; the smaller stomachs can only suck in their booty. They are generally white, with blue dots on the basis.

The descriptions given by most authors are suited only to the genus, but are nowise adapted to distinguish the species from one another. The same may be said of the four descriptions in Lamarck, Histoire Nat. des Animaux sans Vertebres, t. ii. p. 480. Among those seen by us, four species may be distinguished:

A. The gristly shell (when you turn the longest side of the body towards you) extends from the front angle of the right side of the body to the back angle of the left side.

1. *Velella.* — The membrane of the body dark blue; the pretty convex gristly shell is but a very little lighter, the usual transverse line dividing the shell in two, very deeply grooved; the skin surrounding the sail beneath, dark blue; above, paler; the *tentacula*, blue at the basis, at the end reddish-yellow. Length of the body, two inches. — At the Cape of Good Hope.

2. *Velella* — Membrane of the body, dark blue only at the edge; gristle shell, pale yellow; the skin enclosing the sail, colourless: the *tentacula*, at the basis, blue; at the end, reddish yellow. Length of the shell, scarcely an inch and a half. — In the North Pacific Ocean, about the thirtieth degree of latitude. (For the most part, animals not fully grown.)

B. The gristly shell extends from the front angle of the left side of the body, to the back angle of the right side.

3. *Velella.* — Membrane of the body, dark blue; shell brown; the skin enclosing the sail, pale blue; the tentacula, at the basis, light brown; at the end, dark blue; body three inches long, but much narrower in proportion than the two preceding. — Under the equator, in the neighbourhood of Radack.

4. *Velella.* — Membrane of the body, dark blue, shell yellow; the skin enclosing the sail, remarkably green. Length above two inches; nearly an inch and a half broad. — In the Northern Pacific Ocean, north of the Sandwich islands.

III. PORPITÆ.

THE body of the *Porpita* is circular, likewise incloses a very thin transparent gristle, which is here also circular, furnished with many (forty-two) raised radii, and several concentric rings. The exterior skin rises above the gristle, as a narrow rim. There is no trace of a sail. On the under-superficies, you observe in the middle, a large stomach, surrounded by innumerable smaller ones, exactly as in the *velella;* there are, likewise, small *tentacula* attached to the edge of the gristle. A peculiarity of this species is, however, their very long, club-shaped, triangular seizers, which are likewise inserted in the edge of the gristle, of different lengths and thicknesses, (according to the age,) and provided, on their edges, with round suckers. These are always placed at some distance from each other, with respect to the length of the seizer; the suckers, however, of the three different borders stand opposite each other; the seizers are not regularly triangular, but one of the superficies is much smaller than the two others.

The *Porpita* seems destined to fish only on the surface of the sea, as it stretches out its seizers in an horizontal direction, like radii.

The species frequently observed by us in the North Pacific Ocean, from the equator to the 40th degree of north latitude, has a dark violet gristly shell; the skinny edge was blue; the stomachs white, with blueish spots on the basis; the *tentacula* dark blue; the seizers a very little greenish, almost transparent. The disk of the body measured, in diameter, one inch, the seizers nearly two.

I must here repeat what I said respecting the difference of the species of the *velella*. *Porpita nuda*, Lam. Hist. Nat. des Anim. s. Vertebr. tom. ii. p. 484. n. 1. is probably an individual with its seizers torn off, such as we also have frequently caught, because their seizers easily remain hanging in the net, and tear off. — *Porpita appendiculata*, Lam. l. c. n. 2. is a common *Porpita*, mutilated by small crabs, of the exterior skin of the edge of which only three fragments remained; we likewise met with such.

Porpita glandifera, Lam. l. c. n. 3. gives the description of an uninjured individual; also the *Porpita gigantea*, Lam. l. c. n. 4. which (if it is really so) can be distinguished from the preceding by the blue colour of the seizers.

In conclusion, I will observe, that the *Physalia* are furnished, besides the suckers, only with arms for seizing their prey; the *Velella* only with tentacula, but the *Porpita* with both.

DESCRIPTION

OF

NEW FOREIGN BUTTERFLIES,

BY

FREDERICK ESCHSCHOLTZ.

BEFORE I enter into the description of the single species, I must previously observe, that their number is not large, because the principal objects of the expedition were, on the one hand, the north; on the other, the small Coral islands in the South Sea. Besides, we visited the southern climates only in the winter months, consequently, at the season most unfavourable for insects. Hence it was, that Chili and California furnished each but two butterflies. I did not see one butterfly at Oonalashka, during our tolerably long stay, at the three different seasons in which we visited it. The richest harvest, in this respect, was in Manilla, though the inhabitants assured us that it was precisely the most unfavourable time for insects.

I must also observe, that in naming the species,

I have, in this respect, followed the example of Latreille; (as he describes the South American butterflies brought by Humboldt and Bompland,) that is, I have given to the butterflies the names of men of more modern times, as those of ancient history and mythology are of no value; because, for example, other people have got among the Trojan knights, (as Renmus.) Besides this, Latreille was not the first in this respect, as Fabrizius had already adopted the names of Portlandia, Huntera, Allionia, Blomfieldia, Banksia, Solandra, &c. Latreille gives the names unchanged; but I give a proper termination, according to the example of Fabrizius, to distinguish the new from the old.

I. PAPILIO RURICKIA.

P. E. A. alis caudatis fuscis: anticis macula alba quadrata ad marginem posticum; posticis maculis bifidis marginalibus, in utraque pagina miniatis.

From Brazil, from the continent opposite to the island of St. Catharine. — The length, from wing's end to wing's end, is three inches and four lines. The colour of the wings is a dark brown, with a faint greenish lustre; the lower wings are a little darker above than the upper ones. On the back margin of the upper wings there is in the middle an almost square white, yellowish spot, which extends from the back edge, to the second wing rib; and is separated near the edge, from the first broad black tendon. Each of the lower wings has four vermilion spots; the first is broad, nearly lunule at the abdominal corner; the three others consist each of two conical patches, united at the basis, the points of which are turned towards the margin; they occupy the three scallops following the back angle, and always lie between two tendons. In the other scallops, some red dots are observed as an indication of these spots. In the hollows of the scallops, you see, as usual, white scales. The winged tail is half an inch long. Beneath, the wings have a brown silky lustre; the upper wings have, quite at the root, a spot, and a short brown streak under the former, of a vermilion colour. The square white spot is also here. At the root of the inferior wings, we also meet, underneath, with three red spots; from the lowest of them, a long red stripe extends to the abdominal corner; it is a line distant from the inner edge. The four red spots described on the surface of the under-wings are also found beneath, of the same form, but almost of a rose-colour; the three other scallops are here furnished with red spots, divided into two parts, which are, however, a little more obscure than the four others. The tail, on the under side, has also a red stripe. The feelers, legs, and upper side of the body have a black colour; underneath, the body is brown. The edges of the single joints of the feelers project so much, that they seem to be annulated. My specimen is a female.

II. PAPILIO KOTZEBUEA.

P. E. T. alis candatis atris: posticis subtus maculis senis marginalibus, fasciaque transversa abbreviata anguli ani purpureis; fronte, colli thoracis abdominisque lateribus coccineis.

From Manilla.—Some are smaller than the preceding, others larger. The colour of the wings is black, particularly dark on the under wings. On the upper wings of a smaller species, are four rather imperceptible grey yellow stripes towards the back corner, which run parallel with the back edge. In the greater one there are scarcely any traces of them. The under wings are distinguished by their narrowness, and by their broad tail, which is rounded at the end, and half an inch long. On the back angle on the upper side, there is a longish somewhat imperceptible red spot; in the other scallops, only single red dots stand together. But on the under side, the six more or less kidney-shaped purple coloured spots are very clear; they stand around the whole exterior border, placed always between two tendons. A broad band of the same colour extends from the back angle as far as the middle of the wing, to the third tendon, and is intersected by two black tendons running through it. The upper wings, on the under side, have many grey yellow streaks, of which two always run between two tendons, from the middle tendon to the exterior margin. The very small palpi and the woolly forehead are a pale red. The sides of the neck and thorax, as the grooves of the abdomen on the sides, are vermilion; the two last annuli of the abdomen are quite red. I have seen only three females.

III. PAPILIO CHAMISSONIA.

P. E. T. alis caudatis fuscis, omnibus utrinque fascia angusta alba transversa, posticis utrinque maculis octo marginalibus coccineis.

From Brazil.—Its length is three inches four lines from wing's end to wing's end. The colour of the wings and body is a dark green glossy brown; the under wings become quite black towards the margin. Three lines distant from the exterior edge of the upper wings runs parallel with it a yellowish white, quite straight band, scarcely half a line in breadth, which increases a little in breadth towards the back edge of the wing, and is very plain on the disks of both wings. A similar band of the same colour extends also over the under wings, likewise running parallel with the exterior edge, but distant from it about six lines; it becomes broader in the middle, and thus extended, terminates at the innermost tendon. This band is equally strongly marked on both sides. Opposite to each scallop of the under wings is a vermilion transverse spot (not unlike a wine-glass); a similar one stands on the back angle, so that there are in all eight patches. These red spots are particularly visible on the under side of the wing; on the upper side the two top ones are almost entirely lost. The wing-tail is pretty narrow, and only three lines long. The hairs on the forehead, the back, and the thorax, as well as the feelers and legs, are quite black. The palpi, many spots on the breast, the shanks, and the sides of the body have a vermilion colour.

A remarkable peculiarity in this butterfly is the rolled interior edges of the lower wings, as they are covered inside with small grey scales, between which, on the whole length of the wing, grows a white wool, the fibres of which are a line in length. The same is found also in the *Pap. Vertumnus*, F. and *P. Sesostris*, Cram. It has a great similarity with two already-described butterflies, which besides have the same country with it; for which reason I must give the differences. The one is *Pap. Agavus*, Drury's Illustr. of Exot. Ins. vol. iii. tab. ix. fig. 4. It has great resemblance with ours, by its white band and red spots of the lower wings, but differs much from it by the two tailed wings; besides, ours has no large squarish spots. Less dissimilar from it is the other, *Pap. Thyastes*, (Fabr. E. S. n. 77.) as it has several white spots on the surface of the lower wings. My specimen is a male.

IV. PAPILIO RUMANZOVIA.

P. E. A. alis ecaudatis atris; anticis utrinque posti-

cisque subtus basi rubris; posticis medio albis, margine præsertim subtus rubro maculatis.

From Manilla.—Its length from wing's end to wing's end is five inches four lines. Its chief colour is a dark pure black. The upper wings have the usual white grey radii, which are here particularly light; there are likewise from three to four white stripes lengthwise, between the spaces formed by the two chief veins. Almost at the root of the wings is a vermilion transverse spot, which is near four lines long, and one and a half broad. In the middle of the under superficies of these wings the white stripes almost entirely cover the black ground as they greatly increase in breadth. There is here also a red spot at the root, but it lies almost lengthwise of the wing, extending merely between the two chief tendons, and ending in the middle in a roundish elongation. The lower wings are of a very dark black, have in the middle a large uniform white spot, which is divided into five parts by black veins. In the back angle is a long vermilion patch which incloses a uniform black one; an almost square spot is on the under angle, and is likewise vermilion. In the third space there is the third small red spot. Towards the upper edge, on the exterior corner, there are several small sky-blue scales, which are so situated that they form three short streaks.

On the under side the lower wings are beautifully decorated. The large white middle spot has here also almost the same form, with the exception that the fifth innermost division of it is divided by a red stripe, extending from the spot in the back angle; on the other hand, a round white spot is found in the fifth division. The red colour is here richly distributed. A small part of the wing's base is quite vermilion, and is divided by black veins into five parts. The spot in the back angle, described in the upper surface, extends almost to those at the basis, and in the same manner the spot in the hindermost angle fills up almost the whole space; a similar one is in the third division, at the border The spots on the three following scallops are smaller; the last is hook-shaped. In the uppermost space there is a large spot, pretty narrow at the begininng, running along a great extent of the edge, but which then turns inwards and increases considerably in breadth, so that it reaches to the next vein. In this broad part you observe a large number of sky-blue scales. There is, besides, in the middle of the wing, opposite to the broader end of that last described, a scalloped, not large spot, which is half red, half white. The body, feelers, and legs, are black; the thorax has three transverse rows of white points. This last circumstance, as well as the wings being red at the basis, and the white radii on the upper wings, could almost persuade me to take this butterfly for a Trojan knight. My specimen is a female.

V. PAPILIO KRUSENSTERNIA.

P. E. A. alis ecaudatis atris: posticis supra medio cinereis, subtus margine annulis rubris; omnibus subtus basi rubris.

From Manilla.—Its extended upper wings measure four inches nine lines. The chief colour is black. The upper wings have, towards the exterior edge, only faint streaks, of greenish-grey scales, which are, however, marked stronger on the under side. The hinder half of the inferior wings, with the exception of the border, is thickly covered with greenish-grey scales; it is divided into several parts, by the broad black veins. In the back angle there is a very indistinct, narrow, red semi-circle. The upper as well as the lower wings on the under superficies have as much red at the root, as the preceding butterfly. The lower wings are furnished with many variously shaped red spots, of which some form perfect rings, others indicate this form in simple spots. In the back angle there is a long patch, which runs up and down, along the first vein to a point, and incloses in the middle a heart-shaped black spot. Above this spot, some sky-blue scales shine on a red ground. The second space is quite red from the middle to the edge, but leaves open two black spots, lying one above the other, of which the upper one is larger and cornered, the under almost round. The spot of the third space forms a perfect annulus; the fourth is almost the same, only that it is a little open towards the edge. In the three other divisions there is always a crescent at the edge and another towards the middle, with the points turned towards the margin. The internal half-circle shows some sky-blue scales. The body, feelers, and legs are black; some points in the neck, three transverse rows of points on the breast, on each side, as well as streaks on the thighs and legs, are white. These white spots on the thorax, as well as the same colour of the lower wings, at the root, in the preceding butterfly, make me suppose that both may be of one kind. My specimen is a male.

VI. PAPILIO ASTENOUS, F.

From Manilla.—Of this beautiful large butterfly only the female had hitherto been known; the male, which is much smaller, is however greatly distinguished by the marking of the lower wings, which are likewise fulvous; but the exterior border has only narrow black scallops. Near the scallops there are a number of black scales on a yellow ground, and which form another row of scallops, which gives this part the appearance of being covered with black gauze. The radii on the upper wings of the male are whitish, though they are yellowish in the female. The narrow strong-nerved wings of the male are remarkable, of which the upper wings are mostly found half scaled.

It has also a narrow ring round the neck, of a purple colour. The last great horny joint of the body, by which the males, among the knights, may always be exactly distinguished, is of a white colour; while, on the other hand, the very small last joint of the female is covered with black hair. None of the many males and females which I have seen, had, on the surface of the body, (the roots of the wings,) such red spots as are described and drawn by Jablonsky.

I also caught a variety of the female, which, besides its inconsiderable size, was distinguished by the marking of the lower wings, where the fulvous colour was confined to only a large spot in the middle, as in *Pap. Helena*, L. the orange colour. On the under side, a yellow point is in the broad black border, at each tendon. The red band round the neck is in this one also remarkably broad.

VII. PAPILIO LEDEBOURIA.

P. E. A., alis ecaudatis atris: anticis punctis marginalibus, posticis fascia transversa maculari albis; posticis subtus ad angulum ani lunula lutea.

From Manilla.—This butterfly, with respect to its size and marks, has the greatest resemblance to the *Pap. Pammon*, F., so that I at first hesitated to consider it as a peculiar species, particularly as a variety is mentioned with untailed wings. I have seen seven specimens of this butterfly, which were all untailed. The yellow longish spots, which form the transverse band on both sides of the inferior wings, are thrice as long as they are broad; they fill up the space which, in the *P. Pammon*, (in Jablonsky's figure, tab. xix. fig. 4.) is covered by blue scales. On the under side of the lower wings, in the back angle, there is a brownish-yellow spot in the form of a crescent; the convex side is turned to the inner margin. In some specimens a small red-brown point is left also on the upper side of this spot. Besides the usual white spots of the exterior edge of the lower wing, between the scallops, there is on the under superficies, opposite to each of them, another small white transverse spot, sometimes in the form of a crescent. The circumstance of my once having seen this butterfly, in conjunction with the *Pap. Polytes*, F. induced me to examine my specimens, and I found that all three of the present butterfly were males, and my two polytes females. Both butterflies are greatly distinguished from each other, yet they have, on close examination, something in common, but only on the under superficies. Here both have, in common, as well the fine white longitudinal stripes at the roots of all the wings, as the stripes on the belly, and the white points on the thorax.

VIII. VANESSA TAMEAMEA.

V. alis basi fuscis, apice atris: anticis fascia media,

posticis antemarginali fulva, nigro punctata ; posticis subtus obscure viridibus, vitta lata rosacea.

From Woahoo, one of the Sandwich islands.—This butterfly has much resemblance to the *Van. Atlanta*, L., but particularly to the *Atlanta indica* (which is much distinguished as a peculiar kind.) This one, however, of the Sandwich islands, exceeds both considerably (a fourth part) in size; the wings are of a brown colour on the upper side, from the roots to the vermilion bands. The band of the upper wings is three lines broad, has a longish black spot; nearly on the front edge, towards the back edge of the wing, on the interior edge of the band is a large round black spot; the space, also, between the two spots where the first branch diverges from the large middle tendon is black. In the black space, between the band and the tip of the wing, there are three white spots, of which the largest long one is on the front edge, nearer to the band; the smallest likewise on the front edge, but nearer the tip, and the third spot, of a kidney shape, is nearly in the middle. The band of the lower wings lies a line and a half from the exterior black edge, and extends over four membranes, namely, over the third to the sixth. The spots of the intervals, nearest to the internal edge, are only half as long as the two others, and are each marked with a large black patch; the third longest spot has also a dot, though very small. At the edge of the third space is a small blue spot, on a brown ground, surrounded by a black ring. On the under superficies, besides the upper spot in the red band of the upper wings, there is another black spot, which is in the triangular space of the band that is prolonged to the root of the wing. Between the large white patch and the band there is a blueish stripe. The space between the two large white patches, to the tip, is covered with scales of a dirty green. The lower wings, upon the whole, are, on the under side, a dirty green, with darker scalloped bands. From the middle of the front edge to about the middle of the wing extends a white greenish scalloped patch. The place on the upper side, where the red band lies, is here of a pale rose colour, but which has a dirty look, on account of the many green scales spread over it. The body on the upper side is brown, covered with greenish-grey hair beneath, as well as the palpi and legs; the feelers black, with white clubbed ends, eyes brown.

Tamaahmaah, the great hero, and first king of all the Sandwich islands.

IX. CYNTHIA ROESELIA.

C. alis ferrugineis, fascia lata rubra communi, anticis fascia altera alba.

From Brazil.—Of the size of the *Pap. urticæ*, L., only the lower wings are narrow, elongated, and the corner of the exterior edge so much protruded, that the wings might almost be called tailed. The ground colour of the wings, on the surface, is blackish-brown; a vermilion band, from two to three lines in breadth, runs transversely over the superior wings, continues lengthwise over the lower wings, and terminates in a point before the back edge. In the band of the upper wings there are, in front, two indented black stripes; a similar one runs likewise along the under wings, lengthwise of the band. Close to the band of the superior wings, after a narrow brown space, follows a blueish white transverse band, consisting of five longish spots, of which the two hind ones have an indenture outwards. Besides these, there is at some distance from the exterior edge of all the wings, a row of small white spots, which become gradually narrower and more obscure on the lower wings. Of the under side, it is only to be observed that all the dark colours of the surface are much paler, particularly the red band of the under wings, is almost uniform in colour with the whole wing; there is, however, in the middle of it, a small brownish spot. The body at the top is brown, the bottom grey; the feelers are brown with yellow clubs.

X. APATURA RARICK.

Ap. alis ferrugineis: anticis fascia abbreviata, posticis disco albis, anticis macula magna lutea ad marginem posticum.

From the Coral islands of Radack, in the

Northern Pacific Ocean. — Of the size of the *Ap. Bolina.* The ground-colour is dark brown. The superior wings have a short white band, consisting of four longish spots, and which takes its direction from the middle of the front edge to the exterior border ; it extends over four membranes, from the second to the fourth. At a small distance from the exterior margin there are six other spots, of which the upper one is very large. A broad, crooked, brownish, yellow patch extends from nearly the middle of the wing, to the back edge near the back angle. The upper half of the under wings has somewhat of a brownish violet lustre. In the middle of the wing there is a large longish patch, the under and exterior edge of which is bordered with a dirty yellow ; on the under side of the wings there is the same distribution of colour, only that the darker colours are here paler, and more indistinct ; the superior wings have, at the fore-edge, between the roots and white band, three white spots encircled with black. Besides this all the wings at the external edge are encircled with a double row of longish grey spots. The body at the top is dark brown, underneath spotted white ; the feelers black, with yellow tips ; the palpi and feet whitish-grey.

Rarick is one of the chiefs of Romanzoff's group, who first approached us with tokens of friendship.

14

XI. APATURA KRAIMOKU.

Ap. alis fuscis, marginibus duplici serie punctorum alborum.

From Guahon, the most southern of the Mariana islands. — Of the size of the *Ap. Bolina*, of uniform dark brown, with an olive green lustre. Close to the external edge runs a row of small whitish-grey dots over all the wings, of which two always stand in one membrane. At a distance of two lines from the edge there is a second row of larger white spots, of which there is always one in a membrane. On the under side, besides the marks of the upper side, there are two white spots at the front edge of the superior wings, near the roots. Body and feet brown, feelers black.

Kareimoku, governor of the island of Woahoo, one of the Sandwich islands.

XII. IDEA ABIGAR.

I. alis albis nigro cinctis, margine punctis albis duplici serie, anticis fascia apicali transversa alba, medio posticisque subtus striis ferrugineis.

From Manilla. — This species is of the size of the *Pontia Cratægi*, only the wings are narrower. The chief colour is white, but the upper wings are marked with so much black brown colour, that only a triangular space, divided by a black tendon, in two spots lying above each other, is left. There is, besides, a broad white transverse band towards the tip of the wings, to which a large cornered point, near the exterior edge, is joined. The exterior black edge is marked with a double row of white dots, but which vanish near the tip. At the roots of the wings there are two straight yellow-brown stripes; and on the exterior edge of the large white spot, a third crooked one. The under wings are likewise white; their foremost edge, bordered by a narrower, the hinder edge, however, with a broader brown band. Near to the border, in the band, there is a double row of white dots, of which four are always placed in a membrane. In the third membrane there is a hard, black, longish mass, which indicates the Sack (*Sach*) of the relative species in this place. On the under side, the superior wings are marked the same as on the upper side, only the dark brown is here more reddish brown. The lower wings have here, in the broad brown border, many longish light brown spots, which encroach at the back edge with their points upon the white ground. Scutcheon, feelers, and legs are black; the abdomen brown above and yellow beneath. White dots are on the forehead, head, the roots of the wings, above and below, and the legs, and on the basis of the lower wings, even on the under side; the stripes on the thigh and palpi white.

Abigar, chief of a small island of Romanzoff's group.

XIII. IDEA MANUJA.

I. alis rufo ferrugineis, nigro cinctis, punctis duplici serie marginis, maculisque utrinque lacteis, posticis subtus nigro venosis.

From Brazil. — This butterfly has great similarity with the *J. Eresimus*, with respect to size and marking; in the latter respect, also, with *J. Berenice*. Its colour is reddish brown; the front and exterior edge of all the wings is bordered black; particularly broad on the lower wings. On the same there is, as usual, a double row of white dots, of which those of the outer row vanish on the tip of the superior wings; also, towards the back angle a spot is wanting in each row. On the front edge of the upper wings are three; then again three others, close to these two; and, lastly, along the external border, four larger milk-white spots. In the front part of the under-wings there are four white spots. On the under side the wings are marked the same as on the upper, only the double row of larger dots is here interrupted. The lower wings have here a somewhat lighter colour; all the veins are black, some of them are edged with a narrow white margin, as the middle points of the upper side are here longer, and follow the veins for a little way. Head, scutcheon, thorax, feelers, and palpi are black; head, shoulders, and thorax have white spots. The colour of the abdomen is dark brown underneath, with a white line lengthwise.

Manuja, a person of rank in the Sandwich islands, who made the voyage from Owhyee to Woahoo, in the Rurick.

XIV. IDEA PLEXIPPUS, L.

Var. alis posticis margine nigro supra punctis albis duplici serie.

Of this butterfly, which is said to be also met with in America and China, I caught a variety in California, which has also the double row of white dots on the black border of the lower wings, on the upper side; here also the double row of dots on the disk of the upper wings towards the back angle is not interrupted, as in Herbst's drawing. Of the black spot of the lower wings on the second tendon (from the inside), mentioned by Herbst, I see no more traces in my specimen than in Herbst's drawing itself.

XV. EUPLOEA KADU.

E. alis supra atro cœruleis: anticis macula albo vittata punctisque marginis cyaneis; posticis macula flavescente.

From Guahon, the most southern of the Mariana islands. — This butterfly has entirely the form, and almost the marks of the *Pap. Phænareta* (Schaller, Naturforsch, 21. St., s. 177. tab. v.) Its superior wings are particularly distinguished by the great convexity of the back edge, so that they always cover half of the under wings in their natural extent. The colour of the upper wings at the top is dark blue; in the middle, between the first and second tendon from the back edge, is a large heart-shaped spot, of a beautiful ultra-marine colour; the points are turned towards the roots of the wings. In the upper part of this patch there is a long, blueish-white transverse spot. Above the large heart-shaped spot, there is another on the upper edge, and on the external margin five other azure blue dots; near the tip of the wing there are two white dots. The lower wings have two chief colours, which are separated from each other by a line running from the root to the middle of the exterior edge. The front half looks faded, as it is of a glossy, yellowish-grey colour, which becomes dark grey towards the middle. The back half, on the other hand, has a cast of blue on a dark brown colour. In the front half, there is in the middle a large longish spot of a yellow brown, which looks mealy, and is contiguous to the dark brown colour. In the middle of the exterior edge there are three other azure spots, of which the front one is the largest. The under side of the wings shows a dark brown colour. Quite close to the exterior border of the superior wings there is a row of milk-white dots; a second, likewise milk-white row, answers to those described on the upper side, only those nearest to the tip of the wing are smaller. Between the second and third tendon from the back edge, there is a transverse, oval, milk-white spot in the middle. The space from the second tendon to the exterior edge is of the same colour as the front part of the lower wings on the upper side; there is here, likewise, in the middle, a similar mealy, yellowish-brown spot. There is besides, at the upper edge of the spot last described, a black-brown oval ring, which answers to the white stripes in the large heart-shaped spot on the upper side. The dark brown colour of the under wings shines a little reddish. Quite close to the back edge are eight white dots; another row, a little farther distant from the front and back edge, consists of nine blueish dots. Body, feelers, and legs, are dark brown. On the head, shoulders, thorax, and roots of the wings on the under side, there are the usual white spots.

Kadu, a native of the Carolinas, remained eight months on board the Rurick.

XVI. PAPHIA BŒBERA.

P. alis glaucescentibus supra late nigro cinctis, atroque venosis, posticis subtus margaritaceis.

From Mañilla.—Of the size of the *P. Cratægi*, L. The greatest part of the wings on the upper side is pale sea-green. The sector edge of the superior wings is bordered with a black band, one line and a half broad; the exterior edge of the same wing with one three lines broad; and, lastly, the external edge of the lower wings with one of a black brown colour four lines broad. All the tendons are broad, and of a black colour. The dark edge of the under wings is glossy at the sides, but faint at the back angle. On the under side, the superior wings have a greenish colour like mother-of-pearl, the middle tendons are bordered with dark brown, but the tip and exterior margin of the wing is grey. The lower wings are covered here and there with a lustre resembling mother-of-pearl; the tendons and several places of the external edge are of a very pale grey brown colour. The colour of the upper part of the body is properly black, but covered with sea-green scales and hair; beneath, on the contrary, of a white colour. Feelers black; eyes brown; feet white.

XVII. NEPTIS ILLIGERA.

N. alis atris: anticis stria longitudinali maculaque rotunda, posticisque fascia transversa albis.

From Manilla. — Equal to the *N. Melicerta* and *Agathe* in size, and similar in the marks. The colour of the wings is black brown; a white greyish stripe extends from the root of the superior wings to about the middle of them; between the internal end and the back angle of the wings is a large roundish white spot; three distinguished but very small white dots lie in a row between the large patch and the tip of the wing. Before and behind these three there are two other white dots. On the exterior edge run two very indistinct light brown lines, of which the interior one is abbreviated.

Quite close to the external edge in the middle, a longish white spot is distinguishable, which joins the two white transverse bands of the lower wings, when the upper ones are naturally spread out. This band occupies the middle of the wing, and extends from the internal almost to the sector edge; a light brown line surrounds the exterior edge at some distance from it. There is more white on the under side; the longitudinal stripe of the upper wings has become broader, and there is another narrow stripe also lengthwise above it. The exterior edge is almost surrounded with a treble row of lines, but which are frequently interrupted. On the lower wings there is, above the broad band, a narrow one of equal length, a little crooked. The exterior edge is adorned with a treble white border, of which the middle line is the broadest. The body, at the top, is of a black-brown colour in common with the feelers; at the bottom it is white, as also the palpi and thighs; the other part of the legs is brown. There is besides, on the front edge of the back, a spot in the form of a horse-shoe, sparkling with green and red gold colours.

On comparing this species with the *Melicerta*, it will easily be seen that the present is distinguished by the round spot and narrow streak of the superior wings, and by wanting the white margin of the lower wings on the upper side.

XVIII. ACRÆA CLAUDINA.

Acr. alis flavis, supra extus nigro cinctis, anticisque fascia nigra transversa anteapicali, subtus fusco irroratis.

From Brazil. — The colour of the wings on the upper side is yellow-ochre. The size is somewhat smaller than the following, *Ac. Sirena;* the back angle of the present species is also very cornered, by which it deviates from the rest of this genus. A broad part of the tip and exterior edge of the superior wings is black, as also a narrow band, which is however very broad in front, extending from the middle of the front border, towards the back angle, to the external edge. The middle of this transverse band unites by an oblique black line with the black tip of the wing. Between the roots of the wings and the transverse band, there are, at the front edge, four small scalloped transverse streaks, which have a resemblance with those of the species *Argynnis*, F. On the lower wings there is a black exterior edge, which is thickened towards the back angle, by two fine black scalloped streaks, and some black scalloped stripes at the root. On the under side the chief colour is light yellow. The transverse band of the superior wings is here brown and indistinct. Two fine scalloped brown lines surround the exterior edge of all the wings; similar lines are seen at the root of the superior ones, and the front half of the lower wings, in transverse lines; and on the latter, ranged in single lines. There is also in the middle of the lower wings a larger, and a row of smaller brown dots bordering on the scalloped line of the edge. Body and feelers at the top dark brown; beneath, both white; the clubs of the feelers, and the legs light brown.

XIX. MECHANITIS MERIANA.

M. alis atris: macula baseos (posticorum antica) fasciaque externa fulvis, supra media, subtus undique albo punctatis.

From Brazil (Rio Janeiro).—The size and form coincide with the *M. Euterpe*, to which it is very similar; but according to a mere description, it is difficult to distinguish it from the *Susannah*. The chief colour of the wings is black; on the superior wings you first observe a triangular reddish yellow spot, which extends from the root to the middle of the wing, is here rounded off, and leaves behind a narrow black rim on the edge of the upper and lower wings. Between the tip of the wing and the large patch runs a broad scalloped reddish-yellow transverse band, on the inner side of which deep indentures are observed. It comes very close to the sector edge, but it keeps very distant from the exterior. Between the two spots there are four scattered white dots. From the root of the lower wings to the middle of them, a longish quadrangular reddish-yellow spot extends close to the front edge, which has two indentures towards the exterior edge. There is a broad scalloped transverse band of the same colour, a line and a half from the external margin, which reaches neither the front nor the back edge; inwards it has indentures, outwards projecting scallops. On the interior border of the wing between the root and band there are five white points, of which four form a row; other indistinct ones are between the two patches. On the under side we must observe the greatest number and distinctness of the white dots in the middle, and a row of dots on the edge of all the wings. Besides these, the reddish spot at the root of the lower wings is here divided into a small spot near to the root, and a larger one in the shape of a cross. Of the parts of the body I can only observe of the back, breast, and legs, that they have white dots; the rest is wanting in my specimen.

XX. PONTIA HENNINGIA.

P. alis atris: anticis fascia lata transversa alba, posticis supra medio maculisque subtus lateralibus sulphureis, basique coccineis.

From Manilla.— It is a little smaller than Herbst's drawing of *Pasithoe*, with which it has great similarity. From the outer edge of the upper wings, not far from the root, begins a white band three lines broad, and extends towards the back angle, which it does not however reach. The root of the wings is made grey by single white scales; without the band, and near it, is a white spot. Towards the exterior edge and tip are several indistinct grey spots. The under wings, quite at the root, are blackish-grey, then a little white-grey, and after this follows the sulphur colour, which is half-enclosed by a black band, which occupies the greater part of the front, and the whole exterior edge; the middle tendons are also black. On the under side, the grey spots on the tip of the superior wings are transformed into five white ones. In the lower wings, besides the yellow colour, there are three longish spots at the exterior edge, and a round one in the middle. The root of the wings is here scarlet. The back part of the body white, the other parts black grey.

XXI. PONTIA OLGA.

P. alis anticis albis, venis margineque nigris, subtus apice flavo maculatis, posticis aurantiacis, margine nigris.

From Manilla. — This species is of the size of the *P. Rapæ*, and has the greatest similarity with *Judith*. The upper wings are white; the last end of the sector edge, the whole exterior edge, and the front tendons, are black. The under wings have a gold yellow colour; the front and external edge are bordered with black. On the under side the black colour of the upper side has become brown; the two middle tendons are coloured broader, and on the tip there are three light yellow spots. The lower wings are here of a sulphur colour; only the exterior edge is brown like the three foremost tendons. Feelers, blackish brown, scarcely visible on the inner side, with white dots half way. Back brown, with white hair; thorax white; abdomen at the top yellow; underneath white; feet brown.

P. Judith, F. is distinguished by the red spotted edge of the lower wings, and by the want of the three yellow spots on the under side of the tip of the upper wings.

XXII. PONTIA MERCEDIS.

P. alis albis, macula antica triangulisque duplici serie marginis nigris, posticis subtus flavis, albo venosis, antice fulvo cinctes.

From Talcaguano in Chili. — Nearly as large as the male of *P. Brassicæ.* The colour of the wings on the upper side is white, and a very strong lustre of mother of pearl at the root. A long quadrangular spot extends from the middle of the sector edge of the superior wings, as far as the third of the breadth of the wing. About three lines from the external edge, there is a row of triangular spots, with their points turned to the exterior edge, the corners of which, at the basis, are hooked. The one nearest to the back edge is only a longish spot. All these spots are in the membranes, between the tendons. From the exterior edge itself, there begin club-formed spots, but which lie on the tendons with their basis, and run almost to the other row. On the lower wings there is likewise this double row of spots, but they are fainter, and those on the edge divided by the white tendon. The upper wings are also white underneath, but a broad space of the tip is of a light yellow colour. The large spot is here, like the double row of the smaller ones, much fainter, and does not extend to the sector edge. The lower wings are here light yellow; the front edge of a narrow gold yellow; the white tendons encompassed with a narrow blackish line. The double row of triangular spots is nearly the same as on the upper side; in the middle of the wings there is a uniform white spot. The body at the top is blackish, underneath (as also the legs) white. The feelers black, with white annuli.

XXIII. COLIAS ANDREA.

C. alis albis margine antico externóque late nigro cinctis, posticis subtus sulphureis.

From Manilla. — Of the size of the *C. Rhamni*, and has much resemblance to the *Phryne*, and the female of the *Drusilla* (Herbst Tab.); wings white above; the broad sector and vandyked exterior edge of the superior wings is of a blackish brown; the tip however remains white, and has only black tendons. The exterior edge of the lower wings is broad, and blackish brown. Underneath the upper wings are also white, but the tip has a broad brown edge, and incloses a longish white patch. The colour of the lower wings underneath is of a sulphur yellow; the exterior edge, still broader here, is brown. Body blackish, covered with white scales and hair, feelers blackish-brown, legs and feet yellow-brown.

XXIV. HIPPARCHIA STELLERA.

H. alis supra fuscis, subtus albicantibus, fasciis duabus communibus fuscis, anticis, ocello magno utrinque bipupillato, posticis supra duabus subtus sex.

From Manilla — This species is exactly of the size of the *H. Lisander* (Herbst, tab. 195. fig. 4, 5.) with which and with *Baldus*, F. it has great resemblance. The colour of the wings at the top is brown, with a faint lustre of olive-green; the superior ones have on the usual place a large black eye, with a yellow iris and two small steel-coloured pupils. The lower wings have in the middle of the exterior edge, two eyes, half as large, and with only a single pupil. The exterior edge of both wings is darker. Beneath, the colour of the wings is whitish; the external border, however, as well as two narrow transverse stripes, innumerable small streaks, the rings round the yellow iris, and the front edge of the superior wings, are brown. The body brown, beneath whitish; feelers black with white rings; at the end brownish red. This butterfly has besides two longish transparent bladders at the roots of the superior wings, particularly beneath, which are extended tracheæ and facilitate its flight.

XXV. LYCÆNA SCHÆFFERA.

L. alis nigris, anticis utrinque medio albis : omnibus subtus basi argenteo maculatis, posticis fasciis duabus alibis : postica atro maculata.

From Manilla.—About the size of the *L. Damis*, and *Æsopus*, with which this kind in some measure coincides in form and marks. At the top the wings are blackish brown, and shine in a certain light, like verdigris ; the upper wings have here a longish white spot in the middle, which runs pointed towards the root ; beneath, this patch extends to the back edge. There is on the root or near the sector edge, a broad short stripe of a silvery-green colour. Above, the inferior wings are uniform blackish-brown ; beneath, they have on the basis one large and two small silver-green spots. There are, then, two broad white bands lying above each other, of which the hindermost bears eight longish black spots in a row. The brown colour is lighter on the under side. The upper part of the body, as well as the white annulated feelers, has a blackish-brown colour ; the body beneath, and the feet, are white.

XXVI. LYCÆNA CRAMERA.

L. alis supra brunneis, subtus griseis, fascia communi marginali utrinque fulva, subtus punctis nigris, albo cinctis.

From Teneriffe. — Size and form of the *L.*

Scarus. The colour of the wings on the upper side is dark brown, with a green silky gloss. Scarcely half a line from the exterior edge runs a yellowish red, somewhat scalloped, band across both wings, and both sides, more than half a line broad. Beneath, the common colour of the wings is grey. The superior wings, near the red band, have eight large black spots, standing near each other, which are bordered with white rings; only one of these spots stands by itself; the others form almost a row. On the lower wings, towards the bottom of each, there are twelve very small black dots bordered with white, in three transverse lines; the top one has four, the middle three, and the back one five dots. The edge beyond the red band is white on the lower side, with a row of black dots. Feelers with white rings, and body black above, and underneath grey.

The species described is probably a female; the male may look very different.

XXVII. CASTNIA PALLASIA.

C. alis anticis fusco virentibus, griseo fasciatis, posticis atris, fascia baseos punctisque marginis albis, fascia media maculari rubra.

From Brazil. — It has great resemblance, with respect to size and marks, with *C. Licus.* The ground colour of the superior wings is dark olive-green; on it are observed two spots, and a scalloped band lengthwise, of a brownish-grey colour.

The smallest triangular spot lies on the sector edge over the middle of it; the second, three times as large, on the back edge, near the back angle; the band between both extends from the back edge, nearer the root, towards the point, and from its upper half spreads out a broad branch towards the exterior edge. The lower wings on the basis are brownish-violet; then follows, towards the middle, a broad white transverse band, crossed by brown tendons. The ground-colour of the rest of the back part of the wing is black. On this is a transverse band, consisting of eight vermilion spots, of which those nearest the front edge are the smallest. Exactly on the outer edge are five larger, and two very small white dots. On the under side of the upper wings the ground-colour is of a glossy blue; the band which runs into the great spot and the little spot are here yellowish-white; at the exterior edge, there is a row of seven square whitish-grey spots. At the root of the wings, on the fore-edge, there is a red stripe. The marking of the lower wings on the under side is only so far distinguished from the upper side that the band and spots are here larger and more cornered. The body is brown with a greenish lustre at the top; the last annulus of the body yellowish-red; feelers black; legs with red stripes.

XXVIII. SPHINX PUNGENS.

Sph. alis anticis griseis, nigro fasciatis, annulo

medio striisque duabus longitudinalibus atris, posticis basi rufis, nigro trifasciatis, abdomine maculis lateralibus rufis atrisque alternis.

Sphinx Convolvuli var. Drury Ins. 1. tab. xxv. fig. 4. — From Port Jackson, in New Holland. — This species, connected also with *Sph. Convolvuli* of Fabrizius, has properly more resemblance with *Sph. Ligustri* than with the former. The principal colour of the upper wings is whitish-grey on a brown ground, with many scalloped blackish-brown transverse bands, of which one in the middle is the largest and darkest. Distinguished are a small ring and two short stripes in the middle, and another stripe, with scalloped ends, at the tip of the wing. The lower wings are red at the root, towards the middle pale pink, the deep exterior border is grey. Three black bands are particularly distinguished: one almost annular, inclosing the red colour at the root, a middle narrower scalloped one at the interior edge, and a third broad band on the grey edge. The superior wings have beneath, a brownish-grey colour; two to three lines from the exterior edge runs a narrow wavy blackish-brown transverse line. The lower wings are here light grey, which becomes always whiter towards the inner edge. A black band, particularly broad at the inner edge, runs over the middle of the wing; behind it follows a white band; and, lastly, the wing is bordered by a greyish-brown broad edge. The body is grey above and at the sides; on the back on each side near to the roots of the wings runs a broad black stripe, which, at the beginning of the abdomen, first runs towards the middle, but suddenly turns across to the side. On each side of the abdomen are five square rose-coloured spots, which become smaller towards the end of the body, and at last indistinct. Between the two upper spots there is a short black stripe, between the rest a broad square black spot. Beneath, the middle of the body is white; on the belly are three black dots lengthwise; the three last tracheæ on each side are black. The short feelers are in front dark brown, behind white, legs grey.

Sph. Convolvuli exceeds this species in size; it wants the small annulus and the longitudinal stripes of the upper wings; its under wings are grey, &c.

The present species might much easier be taken for *Sph. Cingulata*, F, which may be distinguished by the white point of the upper wings and the five black spots on the belly.

Our species is said to sting very severely with its proboscis.

XXIX. ZEUZERA VIRIDICANS.

Z. alis flavis, maculis duabus magnis punctisque viridibus.

From Manilla. — Exactly of the size and form of the *Hepialus humuli*, F. The ground-colour of the wings is light yellow, which is covered with many large and small green satiny spots. Of this colour are the roots of the upper wings, then a large round spot, and then follows a long patch; besides this, there are ten smaller spots. On the lower wings there is a transverse band at the exterior edge, consisting of one internal round and one large long spot; at the front edge and in the middle some smaller spots. The underside of the wings is not at all different from the upper side. We have still to notice a brown curved bristle about four lines long on the basis of the lower wings. The black feelers are broadly fringed up to the middle, then narrow in the form of a saw. The hair on the shoulders, head, and thorax is brown; the rest is rubbed off in my specimen.

XXX. NOCTUA BAMBUCINA.

N. alis atris, anticis chalibeo strigosis, apice annuloque collari, posticis margine lato externo anoque aurantiacis.

From Manilla, on the bamboo trees. — It is of the size of the *N. Libatrix.* The upper wings are dark-brown near to the sector edge, and at the back edge is a long stripe of a steel blue colour; of the same colour two points and one hook in the middle; and lastly, at some distance from the exterior edge, eight short streaks above each other, which also form a transverse band. The tip of the upper wings, the broad exterior margin of the under wings, an equally broad part of the end of the body, and lastly, a collar, which extends over the thorax in a conical form, are all of a reddish yellow colour. The other part of the lower wings and the upper part of the body is a blackish violet blue. Underneath, the marks of the wings are nearly the same; the steel-blue streaks are wanting in the upper wings; the wings themselves however, have a bluish lustre in the middle. All the tendons are lighter below, than the ground colour. The black thread-like feelers (not at all jagged) are pointed at the end, and before the end there is one place white. The last limbs of the large black feelers naked, cylindrical; feet blackish grey.

AREOMETRICAL OBSERVATIONS,

FROM THE 18TH OF JULY 1816, TO THE 13TH OF APRIL 1818.

Time. 1816.	Temperature of the Sea Water.	Temperature of the Air.	Weight of the Areometer.	Specific Gravity of the Sea Water.	The Ship's Latitude.	The Ship's Longitude.	Observations.
	Reaum.	Fahrenh.			° ′ ″	° ′ ″	
July 18.	4.8	49.6	1 3 6 7	1.02373	52 49 38 N.	200 23 W.	*From Kamtschatka to Kotzebue Sound.*
19.	5.8	50.9	1 3 5	1.02456	53 42 48	196 55 31	
20.	5.3	54.3	1 3 5	1.02438	55 24 47	194 4 7	Beering's Island.
21.	5.	48.0	1 3 5	1.02428	+56 52 35	+190 8 19	
22.	5.7	52.2	1 3 5 7	1.02503	+58 15 53	+186 15 49	
23.	5.7	49.7	1 3 6 7	1.02403	+59 32 17	+183 3 13	
24.	5.3	48.5	1 3 5	1.02438	+60 58 29	+179 17 1	
25.	5.1	37.9	1 3 5	1.02431	+62 11 17	+174 36 37	
26.	4.1	53.8	1 3 5	1.02397	+63 0 16	+171 46 25	
27.	6.1	52.7	1 3 6	1.02366	+63 15 19	+172 48 9	
28.	6.5	58.0	1 3 6	1.0238	+63 8 10	+173 47 6	Near St. Lawrence Island.
29.	5.4	49.0	1 3 6 7	1.02392	+64 6 3	+173 17 49	
30.	5.4	48.1	1 3 6	1.02342	65 27 41	+171 4 55	
31.	6.4	59.5	1 3 7	1.02327	66 14 2	+169 17 13	
Aug. 1.	8.2	58.9	1 3	1.02338	+66 37 55	+164 47 42	
2.	10.2	68.2	2, 3 4 6	1.01807	66 35 18	162 41 00	
3.	10.2	61.0	1	1.02007	66 14 24	+162 8 50	In Kotzebue Sound.
11.	8.1	58.0	1 5 7	1.02184	+66 15 30	+163 20 17	
13.	8.8	59.0	1 5	1.02159	66 16 39	163 44 0	

Time. 1816.		Temperature of the Sea Water.	Temperature of the Air.	Weight of the Areometer.	Specific Gravity of the Sea Water.	The Ship's Latitude.	The Ship's Longitude.	Observations.
		Reaum.	Fahrenh.			° ′ ″	° ′ ″	
Aug.	14.	8.2	59.0	1 3	1.02338	66 48 47 N.	+163 56 39 W.	*From Kotzebue Sound to St. Lawrence Bay.*
	15.	9.3	60.3	1 3 7	1.02426	66 36 32	+165 6 44	
	16.	6.9	46.0	1 3 6	1.024	+66 39 00	+166 20 56	
	17.	8.	49.5	1 3 7	1.02481	+66 20 24	+167 58 59	
	18.	2.	43.0	1 3 5 7	1 02375	+66 4 5	+170 19 5	
	19.	1 7	44.2	1 3 5 6	1.02435	66 16 17	+170 32 17	
	20.	2.9	50.0	1 3 5 6	1.02463	+65 38 8	+171 29 42	
	30.	3.1	43.5	1 3 5	1.02357	+64 25 59	+171 4 12	*From St. Lawrence Bay to Oonalashka.*
	31.	5.2	45.5	— —	— 421	63 13 3	+169 40 5	
Sept.	1.	6.8	44.5	1 3 6	— 39	+61 28 6	+170 11 40	Near an unknown river on the American coast, which empties itself into the sea.
	2.	4.	46.0	— —	— 294	59 42 15	+172 4 58	
	3.	4.5	44.0	1 3 5 7	— 427	+58 14 46	+172 17 43	
	4.	6.	48.0	— —	— 513	55 43 26	+171 3 36	
	5.	6.1	46.5	— —	— 516	+54 7 45	+169 57 42	
	6.	6.4	49.5	1 3 6	— 377	+54 32 34	+169 33 19	
	15.	5.8	50.0	1 3 6	1.02356	54 24 55	165 9 26	*From Oonalshka to California.*
	16.	6.7	51.0	1 3 5	— 487	54 3 12	+164 42 19	
	17.	8.4	51.5	1 3 5 7	— 567	52 32 46	164 24 2	
	18.	8.4	55.5	— —	— 567	50 25 1	+161 7 8	
	19.	9.3	56.0	1 3 5	— 576	48 20 16	158 22 14	
	20.	10.4	54.0	— —	— 614	+46 56 30	+156 13 50	
	21.	10.8	56.0	— —	— 627	45 47 3	+153 24 38	
	22.	55.0	12.	1 3 5	1.02669	44 29 40	150 26 6	

Time. 1816.		Temperature of the Sea Water.	Temperature of the Air.	Weight of the Areometer.	Specific Gravity of the Sea Water.	The Ship's Latitude.	The Ship's Longitude.	Observations.
		Reaum.	Fahrenh.			° ′ ″	° ′ ″	
Sept.	23.	13.	59.1	1 3 6 7	1.02653	43 13 36 N.	147 7 49 W.	
	24.	13.2	60.5	1 3 5 7	— 76	42 20 54	143 52 7	
	25.	13.3	58.2	1 3 5	— 7.3	41 38 50	140 27 57	
	26.	13.2	63.0	1 3 5 7	— 76	41 26 12	139 57 57	
	27.	13.3	64.2	— —	— 763	40 52 56	+136 48 45	
	28.	13.2	63.0	1 3 5	— 71	40 16 59	134 8 56	
	29.	13.3	64.4	— —	— 713	39 39 39	131 35 42	
	30.	13.6	59.6	1 3 6 7	— 673	+39 10 00	128 20 36	
Oct.	1.	12.4	58.7	1 3 5	— 683	38 8 12	124 40 17	
	2.	12.	59.2	1 3 5 7	— 719	37 51 13	122 49 47	
Nov.	2.	13.8	60.3	1 3 6	1.0263	35 37 35	+124 57 24	*From California to the Sandwich Islands.*
	3.	14.2	63.0	1 3 6 7	— 704	33 21 11	127 14 29	
	4.	15.2	66.0	1 3 5	— 779	31 44 19	129 25 2	
	5.	15.6	67.9	1 3 5 7	— 842	30 25 14	+130 59 51	
	6.	16.1	67.9	— —	— 859	+28 56 25	+132 52 33	
	7.	16.4	67.9	— —	— 87	+27 49 3	+135 1 3	
	8.	16.8	72.8	— —	— 883	26 46 55	135 26 42	
	9.	17.1	74.2	1 3 5	— 845	26 11 5	135 50 51	
	10.	17.3	72.0	1 3 5 7	— 9	+25 42 16	+136 48 51	
	11.	17.2	68.9	— —	— 897	+25 5 55	+138 1 16	
	12.	17.8	67.9	1 3 5	— 868	24 6 7	139 2 15	
	13.	17.6	66.6	1 3 6 7	— 81	+23 46 12	140 17 24	
	14.	18.	74.9	1 3 5	— 875	22 49 44	+139 19 15	
	15.	17.8	71.1	— —	— 867	23 9 55	139 51 36	

Time. 1816.	Temperature of the Sea Water	Temperature of the Air.	Weight of the Areometer.	Specific Gravity of the Sea Water.	The Ship's Latitude.	The Ship's Longitude.	Observations.
	Reaum.	Fahrenh.			° ′ ″	° ′ ″	
Nov. 16.	18.5	81.0	1 3 5	1.02892	22 34 6 N.	+140 26 49 W.	
17.	18.2	72.5	— —	— 882	22 1 21	142 19 4	
18.	19.	74.0	— —	— 909	21 18 26	145 9 45	
19.	18.9	71.9	— —	— 906	20 38 21	148 3 33	
20.	19.3	77.1	1 3 6 7	— 869	19 50 23	151 00 30	
21.	19.7	76.7	1 3 5	— 983	20 1 19	153 58 14	
22.	19.7	81.1	— —	— 933	20 19 16	+155 15 42	
23.	20.9	79.2	1 3 6	— 873	19 51 59	+155 35 2	Near Owhyee.
24.	20.5	83.8	1 3 6 7	— 91	19 38 16	+155 42 24	
25.	20.4	81.7	— —	— 907	20 5 40	+155 58 44	
26.	20.4	80.8	1 3 6	— 857	21 0 47	+157 2 59	
Dec. 14.	20.	79.9	1 3 5	1.02943	21 9 23	157 59 15	*From the Sandwich Islands to Radack.*
15.	20.	79.2	1 3 6 7	— 893	20 28 34	158 59 19	
16.	20.	80.3	— —	— 893	20 15 55	159 22 57	
17.	20.3	79.9	— —	— 903	19 43 58	160 7 7	
18.	20.1	79.2	1 3 6	— 846	18 14 57	161 42 38	
19.	20.3	79.2	1 3 6 7	— 903	16 42 56	163 52 31	
20.	20.4	79.0	— —	— 907	17 15 17	166 41 3	
21.	20.4	79.9	1 3 6	— 907	16 53 44	168 43 36	
22.	20.5	78.7	— —	— 86	17 3 00	170 1 20	
23.	20.4	81.0	— —	— 857	16 12 49	170 58 12	
24.	20.8	82.7	— —	— 87	14 43 44	173 18 42	
25.	21.1	81.0	— —	— 881	12 49 37	175 24 26	
26.	21.6	86.0	1 3 7	— 848	11 14 30	177 34 38	

Time. 1816-17.		Temperature of the Sea Water.	Temperature of the Air.	Weight of the Areometer.	Specific Gravity of the Sea Water.	The Ship's Latitude.	The Ship's Longitude.	Observations.
		Reaum.	Fahrenh.			° ′ ″	° ′ ″	
Dec.	27.	21.6	82.5	1 3 7	1.02848	11 3 29 N.	179 37 43 W.	
	28.	21.9	84.5	— —	— 859	9 37 14	181 52 41	
	29.	22.	82.1	— —	— 862	9 52 20	184 16 32	
	30.	22.1	80.7	1 3	— 815	9 48 56	186 51 41	
	31.	21.6	78.9	— —	— 798	9 49 57	+ 188 14 42	
1817.								
January	1.	21.7	80.2	1 3 6	1.02902	9 47 40	+ 188 46 22	
	2.	22.1	83.5	1 3 7	— 865	10 5 2	188 50 13	New-year's Island.
	3.	22.	81.9	— —	— 862	9 59 47	189 21 56	
	4.	21.8	81.0	— —	— 855	9 43 21	+ 189 33 45	
	5.	21.8	82.1	1 3	— 805	9 27 55	+ 189 46 34	
	6.	21.8	83.8	1 3 7	— 855	9 30 45	+ 189 39 25	} Inside of Romanzoff's Group.
	12.	21.7	81.4	1 3	— 802	9 28 34	+ 190 0 0	}
Feb.	8.	21.8	83.0	— —	— 805	9 9 49	189 51 14	
	9.	21.6	81.1	— —	— 798	8 53 16	189 26 43	
March	5.	21.2	81.5	1 3 6	— 885	10 27 25	189 58 26	Inside of Krusenstern's Group.
	17.	21.1	76.8	1 3 6	1.02881	13 2 16	190 44 6	*From Radack to Oonalashka.*
	18.	21.2	78.4	— —	— 885	13 45 11	190 38 45	
	19.	20.8	79.3	— —	— 87	14 39 29	190 59 08	Cornwallis Island.
	20.	20.1	79.0	1 3 6 7	— 896	16 38 43	192 13 47	
	21.	19.6	77.8	1 3 5	— 929	17 56 6	193 23 18	
	22.	19.6	77.4	— —	— 929	19 5 36	194 21 22	
	23.	19.6	74.7	— —	— 929	20 15 21	195 5 20	
	24.	19.	76.5	— —	— 909	22 6 2	195 29 56	

Time. 181‾		Temperature of the Sea Water.	Temperature of the Air.	Weight of the Areometer.	Specific Gravity of the Sea Water.	The Ship's Latitude.	The Ship's Longitude.	Observations.
		Reaum.	Fahrenh.			° ′ ″	° ′ ″	
March	25.	18.2	69.9	1 3 5	1.02882	24 16 59 N.	+ 196 3 55 W.	
	26.	17.5	78.0	1 3 5 7	— 907	24 51 9	196 31 00	
	27.	17.4	72.8	1 3 5	— 854	26 24 18	197 19 7	
	28.	14.7	66.4	1 3 5 7	— 811	28 48 0	+ 198 21 48	
	29.	13.5	63.1	— —	— 770	31 17 48	+ 198 54 40	
	30.	13.	59.8	1 3 5 6	— 803	33 0 45	+ 197 41 19	
	31.	13.	57.5	— —	— 803	33 43 38	196 14 34	
April		13.	57.5	— —	— 803	34 23 48	195 13 20	
	2.	11.2	55.0	— —	— 741	34 2 49	194 8 8	On this day the sea acquired its green colour, and retained it.
	3.	11.8	55.3	— —	— 761	34 27 11	193 52 40	
	4.	11.6	58.0	— —	— 754	+ 36 4 55	+ 192 19 6	
	5.	12.1	57.2	— —	— 772	35 35 16	191 49 6	
	6.	11.6	62.2	— —	— 754	+ 37 54 32	+ 190 46 9	
	7.	6.9	55.0	1 3 5 6 7	— 644	+ 40 21 40	+ 189 23 20	
	8.	5.1	46.9	— —	— 581	+ 41 21 44	+ 188 39 32	
	9.	5.8	39.5	1 3 5 6	— 555	41 59 3	186 19 29	
	10.	4.2	41.3	— —	— 501	43 3 0	183 58 56	
	14.	3.8	48.0	1 3 5 6 7	— 537	+ 45 25 8	+ 179 11 53	
	15.	3.1	45.2	1 3 5 6 7	— 513	46 24 22	176 21 10	
	16.	2.3	40.0	— —	— 435	48 22 22	173 38 1	
	17.	2.6	41.0	1 3 5 6	— 445	50 31 20	171 37 36	
	18.	2.8	39.3	— —	— 455	51 42 55	170 12 21	
	19.	2.7	36.9	1 3 5 6	— 449	52 10 50	+ 169 52 26	Near the Aleutian Islands.
	20.	2.7	38.0	— —	— 449	+ 52 20 54	+ 168 38 19	
	21.	2.8	37.	— —	— 455	+ 52 52 2	+ 167 20 38	

Time. 1817.		Temperature of the Sea Water.	Temperature of the Air.	Weight of the Areometer.	Specific Gravity of the Sea Water.	The Ship's Latitude.	The Ship's Longitude.	Observations.
		Reaum.	Fahrenh.			° ′ ″	° ′ ″	
April	22.	2.8	41.3	1 3 5 6	1.02455	53 7 19 N.	+166 5 14 W.	Near the Aleutian Islands.
	23.	2.8	43.1	— —	— —	53 29 9	166 2 21	
	24.	3.1	41.5	1 3 5 7	— 413	53 45 14	+165 38 24	
								From Oonalashka to St. Lawrence's Island and back again.
June	30.	5.3	50.2	1 3 5 7	1.02488	56 4 55	167 53 32	Near St. George's and St. Paul's Islands.
July	1.	4.9	42.3	1 3 5	— 425	+56 41 55	169 7 28	
	2.	4.5	45.0	— —	— 411	+57 5 40	+169 39 25	
	3.	4.7	46.3	— —	— 418	57 4 8	170 1 40	
	4.	5.8	47.0	1 3 5 7	— 506	56 30 32	172 2 37	
	5.	5.8	46.0	1 3 5	— 456	57 2 20	+172 20 18	
	6.	4.2	47.0	— —	— 407	+58 42 50	+171 42 42	
	7.	4.	47.3	— —	— 394	+60 3 50	+171 15 27	
	8.	4.6	46.5	— —	— 414	60 24 10	170 3 15	
	9.	4.	43.9	— —	— 394	61 10 19	170 0 29	
	10.	3.9	47.0	— —	— 391	62 48 47	169 54 51	St. Lawrence Island.
	11.	3.7	47.3	1 3 6 7	— 334	62 59 7	169 36 18	
	12.	4.3	46.5	1 3 6	— 304	62 47 44	169 28 39	
	13.	4.6	46.3	1 3 5	— 414	62 33 16	170 27 9	
	14.	5.1	58.0	— —	— 431	+61 43 6	+171 25 59	
	15.	5.4	49.5	1 3 5 7	— 493	+61 21 48	+171 22 43	
	16.	4.5	47.5	1 3 5	— 411	+60 9 4	+171 46 18	
	17.	5.	46.3	— —	— 428	+59 29 0	+172 16 33	
	18.	5.8	48.0	1 3 5 7	— 506	+58 11 51	+171 14 48	
	19.	4.4	48.2	1 3 5 6	— 508	56 25 35	+169 50 46	St. Paul and St. George.
	20.	6.1	49.9	— —	— 576	55 4 51	+168 17 58	

Time. 1817.	Temperature of the Sea Water.	Temperature of the Air.	Weight of the Areometer.	Specific Gravity of the Sea Water.	The Ship's Latitude.	The Ship's Longitude.	Observations.
	Reaum.	Fahrenh.			° ′ ″	° ′ ″	
July 21.	8.	49.9	1 3 5	1.02531	54 24 2 N.	+ 167 29 16 W.	
August 19.	7.6	55.9	1 3 5	1.02517	54 32 14	+ 166 4 22	*From Oonalashka to the Sandwich Islands.*
20.	5.8	49.9	1 3 5 7	— 506	54 0 42	+ 165 30 47	
21.	7.9	53.0	1 3 5	— 528	54 6 22	164 24 00	
22.	8.	55.0	— —	— 531	52 21 58	+ 164 44 26	
23.	8.2	54.9	— —	— 538	51 2 0	+ 165 49 54	
24.	8.5	54.0	— —	— 549	+ 49 18 36	+ 166 6 1	
25.	9.8	58.3	— —	— 593	48 46 52	166 14 55	
26.	9.4	55.7	— —	— 58	+ 47 48 17	+ 166 23 37	
27.	9.9	58.0	— —	— 597	+ 47 10 20	+ 165 46 21	
28.	·9.9	59.0	— —	— 597	+ 46 38 49	163 43 57	
29.	10.8	59.9	— —	— 627	+ 46 20 46	+ 161 33 3	
30.	11.6	62.0	— —	— 654	+ 45 40 58	+ 159 46 45	
31.	12.4	63.0	1 3 6 7	— 633	46 17 23	156 12 30	
Sept.	12.	62.0	— —	— 619	45 56 52	+ 154 30 6	
2.	11.9	58.8	— —	— 616	+ 45 28 6	+ 155 12 18	
3.	12.5	62.0	— —	— 636	+ 44 35 54	+ 155 12 46	
4.	13.7	65.2	— —	— 677	43 21 3	155 18 28	
5.	14.9	66.0	— —	— 719	42 38 48	+ 154 17 5	
6.	—	68.1	— —	— —	42 1 12	151 34 22	
8.	15.2	69.0	1 3 6	— 679	41 16 34	151 21 50	
9.	15.6	67.3	— —	— 692	40 52 51	+ 150 14 13	
10.	15.8	67.0	1·3 6 7	— 749	40 10 21	147 18 1	
11.	16.	69.7	— —	— 756	38 49 13	147 46 32	

Time. 1817.		Temperature of the Sea Water.	Temperature of the Air.	Weight of the Areometer.	Specific Gravity of the Sea Water.	The Ship's Latitude.	The Ship's Longitude.	Observations.
		Reaum.	Fahrenh.			° ′ ″	° ′ ″	
Sept.	12.	17.6	76.0	1 3 5	1.0286	36 54 39 N.	148 32 10 W.	
	13.	18.4	75.0	— —	— 889	36 8 47	148 9 0	
	14.	18.7	78.9	— —	— 899	35 50 44	147 37 57	
	15.	18.	75.7	— —	— 875	35 28 8	147 47 2	
	16.	18.2	73.7	1 3 6 7	— 832	34 8 58	148 15 55	
	17.	18.8	74.5	— —	— 852	32 3 10	+ 148 54 22	
	18.	19.2	75.6	1 3 5	— 916	+ 29 54 34	150 41 10	
	19.	19.5	78.1	1 3 5 7	— 976	28 51 33	152 10 00	
	20.	19.7	78.5	— —	— 933	28 16 49	+ 152 41 33	
	21.	20.2	77.0	1 3 5	— 95	27 57 47	152 26 42	
	22.	20.7	81.1	— —	— 967	27 50 27	152 21 39	
	23.	20.2	77.0	— —	— 95	26 40 43	152 32 27	
	24.	19.9	78.5	1 3 5 7	— 99	24 13 39	152 50 53	
	25.	20.4	79.9	1 3 6 7	— 907	21 40 20	153 46 18	
	26.	20.5	83.0	1 3 6	— 86	20 13 20	155 2 29	
	27.	21.5	86.5	— —	— 895	20 9 36	155 50 17	
	28.	21.8	84.9	— —	— 905	19 41 31	155 51 20	
	29.	22.	81.9	— —	— 912	20 28 16	156 39 49	
Oct.	14.	20.9	80.8	1 3 5	1.02975	21 10 22	158 13 16	*From the Sandwich Islands to Romanzoff's Group.*
	15.	21.4	82.5	1 3 6 7	— 942	19 57 7	159 41 42	
	16.	21.5	82.5	— —	— 945	19 22 57	161 33 50	
	17.	21.6	82.6	1 3 6	— 898	18 26 18	+ 163 52 56	
	18.	21.7	83.5	— —	— 902	17 33 17	165 39 28	
	19.	21.8	83.1	— —	— 905	16 47 29	167 23 7	

Time. 1817.		Temperature of the Sea Water.	Temperature of the Air.	Weight of the Areometer.	Specific Gravity of the Sea Water.	The Ship's Latitude.	The Ship's Longitude.	Observations.
		Reaum.	Fahrenh.			° ′ ″	° ′ ″	
Oct.	20.	21.9	84.5	1 3 6	1.02908	16 45 12 N.	169 16 37 W.	
	21.	21.8	82.1	— —	— 905	16 0 21	+ 170 52 14	
	22.	22.	84.0	— —	— 912	15 12 2	172 32 10	
	23.	22.2	85.0	— —	— 919	14 31 46	174 13 49	
	24.	22.4	84.8	— —	— 926	13 48 44	176 5 31	
	25.	22.3	84.0	1 3 5 7	— 972	12 48 4	178 37 17	
	26.	22.4	84.7	1 3 6	— 926	11 58 39	180 55 47	
	27.	22.7	85.9	— —	— 936	11 57 55	183 20 36	
	28.	22.8	83.3	— —	— 939	10 30 49	185 34 40	
	29.	22.7	85.5	— —	— 936	9 17 22	188 5 50	
	30.	22.7	78.0	— —	— 936	+ 9 18 18	+ 189 13 26	
Nov.	4.	22.7	82.5	1 3 6	— 936	+ 9 18 59	+ 190 3 39	*From Romanzoff's Group to Guahon.*
	5.	22.8	85.2	— —	— 939	9 48 27	190 36 54	
	6.	22.7	84.3	— —	— 936	9 42 56	191 52 40	
	7.	22.6	83.2	1 3	— 832	9 43 45	193 27 29	
	8.	22.9	85.	— —	— 843	9 40 30	195 7 56	
	9.	22.8	82.7	— —	— 839	9 32 54	197 22 24	
	10.	22.6	81.1	— —	— 832	9 22 11	199 40 29	
	11.	22.8	84.3	1 3 7	— 889	9 19 56	201 25 9	
	12.	22.9	83.4	— —	— 893	9 7 35	202 53 12	
	13.	24.4	85.7	1 4 7	— 845	8 59 05	204 24 30	
	14.	23.1	85.6	1 3	— 849	9 20 59	204 44 26	
	15.	24.3	84.9	— —	— 891	9 25 48	205 0 45	
	16.	22.9	84.0	1 4 7	— 793	9 24 3	205 35 40	

Time. 1817.		Temperature of the Sea Water.	Temperature of the Air.	Weight of the Areometer.	Specific Gravity of the Sea Water.	The Ship's Latitude.	The Ship's Longitude.	Observations.
		Reaum.	Fahrenh.			° ′ ″	° ′ ″	
Nov.	17.	23.1	83.5	1 3	1.02849	10 2 36 N.	206 42 53 W.	
	18.	23.1	83.2	1 3 7	— 899	10 40 59	207 53 2	
	19.	23.	82.7	— —	— 896	11 4 54	209 4 00	
	20.	23.1	84.2	— —	— 899	11 42 29	209 51 20	
	21.	22.8	82.1	— —	— 889	12 28 23	210 53 57	
	22.	22.7	83.2	— —	— 886	13 28 32	212 41 54	
	23.	22.6	83.7	1 3	— 832	13 52 38	214 49 00	
	29.	22.6	83.7	1 3 7	— 832	13 47 2	215 43 15	*From Guahon to Manilla.*
	30.	22.2	83.3	— —	— 869	15 37 4	217 40 11	
December		22.4	82.5	— —	— 876	16 31 34	219 6 36	
	2.	22.	81.2	— —	— 862	17 23 22	220 46 14	
	3.	22.	79.6	— —	— 862	18 24 32	222 53 55	
	4.	21.7	79.8	— —	— 852	19 20 10	225 27 54	
	5.	20.9	79.8	1 3 6	— 875	19 44 48	227 55 44	
	6.	20.9	77.3	— —	— —	19 44 22	229 25 20	
	7.	21.3	76.9	1 3 6 7	— 938	20 00 4	232 12 32	
	8.	20.2	77.0	— —	— 9	20 10 49	235 4 25	
	9.	20.6	77.3	— —	— 913	20 2 11	237 37 55	
	10.	20.7	80.0	1 3 6	— 867	19 12 36	239 43 36	
	11.	22.2	82.0	1 3	— 819	18 00 36	240 7 24	
	12.	21.6	81.7	1 3 7	— 848	16 42 57	240 34 7	
	13.	22.3	84.5	1 3	— 822	15 50 58	240 24 21	
	14.	22.	82.5	1 4 7	— 762	14 44 48	240 2 20	

Time. 1818.	Temperature of the Sea Water.	Temperature of the Air.	Weight of the Areometer.	Specific Gravity of the Sea Water.	The Ship's Latitude.	The Ship's Longitude.	Observations.
	Reaum.	Fahrenh.			° ′ ″	° ′ ″	
Jan. 30.	20.8	80.0	1 3 6	1.0286	13 49 00 N.	241 53 00 W.	*From Manilla to the Cape.*
31.	20.8	79.1	1 3 7	— 281	13 23 28	245 2 55	
Feb. 1.	20.9	79.2	1 3 6	— 875	13 7 44	247 41 21	
2.	21.3	82.3	— —	— 888	11 38 51	249 14 44	
3.	20.7	83.0	— —	— 867	9 41 4	251 0 6	
4.	20.5	83.2	— —	— 86	7 33 23	253 1 16	
5.	21.	78.8	— —	— 878	5 10 00	254 32 23	
6.	21.6	83.0	1 3 7	— 848	2 45 59	255 19 19	
7.	21.9	83.5	— —	— 859	1 19 56	254 23 00	
8.	21.9	83.0	— —	— 859	0 25 51 S.	252 50 0	
9.	22.2	80.6	— —	— 869	1 53 8	252 51 51	
10.	22.3	81.7	1 3	— 822	3 4 0	253 0 22	
11.	23.3	83.3	1 6	— 556	4 21 19	253 25 1	
12.	22.7	82.6	— —	— 536	5 3 23	253 54 00	
13.	22.9	83.6	1 6 7	— 593	+ 5 29 00	+ 254 9 20	
14.	23.3	84.6	1 5	— 656	6 6 41	254 16 30	
15.	23.	85.0	1 5 7	— 696	6 16 56	254 50 45	
16.	23.	82.3	1 3	— 846	6 36 23	255 45 13	
17.	22.7	85.7	1 3 7	— 886	7 19 54	257 10 3	
18.	22.8	85.9	1 3 6	— 939	8 55 29	259 9 21	
19.	22.7	84.2	— —	— 936	10 10 53	261 34 25	
20.	22.3	84.6	— —	— 922	11 42 32	264 25 0	
21.	21.7	83.0	— —	— 902	13 11 37	266 52 0	
22.	21.8	82.5	— —	— 905	14 40 13	269 48 0	
23.	21.8	83.2	1 3 6 7	— 955	16 2 23	272 20 0	

Time. 1818.		Temperature of the Sea Water.	Temperature of the Air.	Weight of the Areometer.	Specific Gravity of the Sea Water.	The Ship's Latitude.	The Ship's Longitude.	Observations.
		Reaum.	Fahrenh.			° ′ ″	° ′ ″	
Feb.	24.	21.3	82.3	1 3 6 7	1.02938	17 14 5 S.	274 58 40 W.	
	25.	21.6	83.5	— —	— 948	18 23 4	277 25 12	
	26.	21.3	82.9	— —	— 938	19 9 35	279 37 0	
	27.	21.8	83.5	1 3 5	1.03005	20 6 57	282 22 18	
	28.	21.5	82.2	— —	1.02995	20 59 31	284 47 47	
March	1.	21.3	81.2	— —	— 988	21 34 43	287 17 52	
	2.	—	81.9	— —	= =	22 2 17	289 40 31	
	3.	21.6	84.1	— —	— —	22 50 53	291 43 37	
	4.	21.3	82.7	1 3 5 7	1.03038	23 7 52	293 26 40	
	5.	—	81.9	1 3 6 7	1.02938	23 59 37	296 34 0	
	6.	—	82.0	1 3 5	— 988	24 31 46	298 55 10	
	7.	20.9	80.5	— —	— 975	25 24 59	302 6 00	
	8.	20.2	79.5	1 3 5 7	1.03	26 8 31	304 18 12	
	9.	20.1	78.0	1 3 5	1.02946	27 4 22	307 18 36	
	10.	19.6	78.7	1 3 5 7	— 979	28 1 57	309 41 00	
	11.	20.1	80.2	1 3 5	— 946	29 5 15	312 25 3	
	12.	20.8	78.0	— —	— 97	29 18 34	313 26 10	
	13.	19.6	77.2	1 3 5 7	— 979	30 22 11	816 33 00	
	14.	18.4	74.7	— —	— 939	31 4 11	318 43 0	
	15.	18.9	77.2	— —	— 956	31 47 38	321 7 41	
	16.	19.1	78.7	— —	— 962	32 21 10	323 13 0	
	21.	18.6	70.4	1. 3 5 6	— 995	31 20 03	325 51 38	
	22.	19.8	73.4	— —	1.03019	32 16 31	327 55 18	
	23.	17.6	72.0	1 3 5 6 7	1.0301	+ 32 50 56	329 27 33	
	24.	17.7	76.1	— —	— 014	33 14 12	330 1 14	

Time. 1818.	Temperature of the Sea Water.	Temperature of the Air.	Weight of the Areometer.	Specific Gravity of the Sea Water.	The Ship's Latitude.	The Ship's Longitude.	Observations.
	Reaum.	Fahrenh.			° ′ ″	° ′ ″	
March 25.	17.4	67.5	1 3 5 6 7	1.03004	33 27 31 S.	330 27 0 W.	
26.	17.4	72.0	1 3 5 6	1.02954	34 1 39	331 48 20	
27.	16.	77.5	1 3 5 6 7	— 956	35 17 50	337 4 18	
28.	15.6	78.9	1 3 5 6	— 892	35 19 32	338 34 16	
29.	16.2	72.0	— —	— 913	34 54 47	339 1 52	
30.	8.9	67.5	— —	— 663	34 23 21	341 45 30	In the Atlantic Ocean.
April 9.	14.2	65.0	1 3 5 6	1.02844	33 21 25	342 39 0	*From the Cape to St. Helena.*
10.	15.8	73.0	— —	— 899	32 54 11	342 47 15	
11.	15.7	65.7	1 3 4 6	— 996	31 57 41	343 58 0	
12.	16.1	74.0	1 3 4 6	— 959	31 13 40	344 45 16	
13.	16.4	67.5	1 3 4 6	1.0302	30 39 02	345 33 30	

TABLE

OF COMPARISON OF THE SPECIFIC GRAVITY OF THE SEA WATER UNDER THE DIFFERENT DEGREES OF LATITUDE AND LONGITUDE, AND AT DIFFERENT TIMES OF THE YEAR.

Latitude.	IN THE NORTHERN HEMISPHERE.																	
Degrees.	Specific Weight of the Sea Water.	Time.	Degrees of Longitude.	Specific Weight of the Sea Water.	Time.	Degrees of Longitude.	Specific Weight of the Sea Water.	Time.	Degrees of Longitude.	Specific Weight of the Sea Water.	Time.	Degrees of Longitude.	Specific Weight of the Sea Water.	Time.	Degrees of Longitude.	Specific Weight of the Sea Water.	Time.	Degrees of Longitude.
		1816.			1816.			1817.			1817.			1817.			1818.	
65	1.0235	July	171	1.0240	Aug.	175	—	—	—	—	—	—	—	—	—	—	—	—
60	1.0242	July	183	1.0242	Sept.	170	1.0241	July	171	—	—	—	—	—	—	—	—	—
55	1.0242	July	194	1.0251	—	167	1.0243	June	167	1.0249	July	168	—	—	—	—	—	—
50	1.0236	June	199	1.0256	—	161	1.0245	April	171	1.0254	Aug.	166	—	—	—	—	—	—
45	1.0253	—	200	1.0265	—	153	1.0253	—	179	1.0263	—	155	—	—	—	—	—	—
40	1.0267	—	200	1.0273	—	133	1.0269	—	189	1.0271	Sept.	147	—	—	—	—	—	—
35	1.0278	—	201	1.0270	Oct.	125	1.0277	—	193	1.0287	—	147	—	—	—	—	—	—
30	1.0277	—	199	1.0285	Nov.	131	1.0279	Mar.	198	1.0292	—	150	—	—	—	—	—	—
25	1.0284	—	197	1.0287	—	138	1.0288	—	196	1.0294	—	153	—	—	—	—	—	—
20	1.0280	May	194	1.0291	—	151	1.0292	—	195	1.0291	Oct.	159	1.0289	Dec.	232	—	—	—
15	1.0280	—	192	1.0288	Dec.	173	1.0288	—	190	1.0291	—	172	1.0286	Nov.	217	1.0284	Jan.	235
10	1.0276	—	188	1.0284	—	181	1.0284	—	189	1.0294	—	189	1.0286	—	206	1.0287	Feb.	251
5	1.0285	—	180	—	—	—	—	—	—	—	—	—	—	—	—	1.0286	—	254
0	1.0285	—	175	—	—	—	—	—	—	—	—	—	—	—	—	1.0285	—	252

TABLE

OF COMPARISON OF THE SPECIFIC GRAVITY OF THE SEA WATER, &c.—*continued*

Latitude.	IN THE SOUTHERN HEMISPHERE.								
Degrees.	Specific Weight of the Sea Water.	Time.	Degrees of Longitude.	Specific Weight of the Sea Water.	Time.	Degrees of Longitude.	Specific Weight of the Sea Water.	Time.	Degrees of Longitude.
		1816.			1818.			1818.	
65	—	—	—	—	—	—	—	—	—
60	—	—	—	—	—	—	—	—	—
55	—	—	—	—	—	—	—	—	—
50	—	—	—	—	—	—	—	—	—
45	—	—	—	—	—	—	—	—	—
40	—	—	—	—	—	—	—	—	—
35	—	—	—	1.0280	Mar.	335	1.0287	April	343
30	—	—	—	1.0298	—	316	1.0299	—	345
25	1.0285	Mar.	113	1.0298	—	302	—	—	—
20	1.0294	April	125	1.0298	Feb.	282	—	—	—
15	1.0289	—	140	1.0292	—	269	—	—	—
10	1.0283	May	157	1.0291	—	261	—	—	—
5	1.0285	—	164	(1.0261)	—	254	—	—	—
0	—	—	—	—	—	—	—	—	—

TEMPERATURE OF THE SEA WATER AT DIFFERENT DEPTHS.

In the Years 1815, 1816, 1817, and 1818.

Days.	Temperature of the Sea Water on the Surface.	Temperature of the Sea Water below the Surface.	Depth in Fathoms.	Temperature of the Air.	The Ship's Latitude.	The Ship's Longitude.	Transparency of the Water in Fathoms.	Observations.
1815.					° ′	° ′		
15. October	+ 68.5	+ 55.7	100	+ 71.1	39 27 N.	12 57 W.	10	
16. —	+ 69.1	55.0	138	72.5	39 4	13 8	10	In the Atlantic Ocean.
—	—	56.0	96	—	—	—	—	
25. —	74.3	56.3	196	74.3	30 12	15 14	11	
1816.								
8. January	54.9	38.8	196	57.6	44 47 S.	57 31	8	Cape Horn.
7. April morn.	78.5	68.5	125	79.2	18 17	124 56	13	
—	—	57.5	175	—	—	—	—	
At noon.	79.6	68.0	125	80.0	—	—	—	
—	—	54.0	200	—	—	—	—	
13. April	80.0	79.0	10	79.8	15 26	133 42	13	In the South Sea.
—	—	79.0	20	—	—	—	—	
—	—	78.8	50	—	—	—	—	
—	—	72.0	100	—	—	—	—	
—	—	56.0	200	—	—	—	—	
12. May	82.5	55.0	300	83.0	1 17 N.	177 5	14	At the Equator.
1. June	74.0	62.0	100	75.0	29 24	199 26	10	
—	—	52.5	300	—	—	—	—	

Days.	Temperature of the Sea Water on the Surface.	Temperature of the Sea Water below the Surface.	Depth in Fathoms.	Temperature of the Air.	The Ship's Latitude.	The Ship's Longitude.	Transparency of the Water in Fathoms.	Observations.
1816.					°	° ′		
6. June	61.0	59.5	10	63.0	37 3 N.	199 17 W.	2	
—	—	56.8	25	—	—	—	—	
—	—	52.7	100	—	—	—	—	
—	—	43.0	300	—	—	—	—	
1817.								
14. September	72.2	72.0	4	75.0	35 51	147 38	11	
—	—	70.9	8	—	—	—	—	
—	—	68.1	15	—	—	—	—	
—	—	57.6	25	—	—	—	—	In the North Pacific Ocean.
—	—	54.0	50	—	—	—	—	
—	—	51.0	100	—	—	—	—	
—	—	42.8	408	—	—	—	—	
13. —	71.9	57.1	25	73.0	36 9	148 9	13	
—	—	52.8	100	—	—	—	—	
—	—	44.0	300	—	—	—	—	
22. —	77.0	75.0	5	76.1	27 50	152 21	16	
—	—	74.5	10	—	—	—	—	
—	—	73.7	25	—	—	—	—	
—	—	67.2	50	—	—	—	—	
—	—	61.0	100	—	—	—	—	
—	—	51.5	200	—	—	—	—	Between Radack and the Marianas.
13. November	87.0	56.2	100	85.0	8 59	204 24	13	
14. —	83.0	77.0	25	84.0	9 20	204 44	—	
15. —	87.4	79.0	15	85.7	9 26	205 1	15	

Days.	Temperature of the Sea Water on the Surface	Temperature of the Sea Water below the Surface.	Depth in Fathoms.	Temperature of the Air.	The Ship's Latitude.	The Ship's Longitude.	Transparency of the Water in Fathoms.	Observations.
1817.					° ′	° ′		
15. November	87.4	59.1	50	85.7	9 26 N.	205 1 W.	15	Between Radack and the Marianas.
—	—	51.4	69	—	—	—	—	
—	—	49.5	101	—	—	—	—	
17. —	84.2	57.4	73	83.5	10 3	206 43		
18. —	83.9	59.9	61	83.2	10 41	207 53		
19. —	83.7	56.6	82	82.7	11 4	209 4		
20. —	84.0	63.0	86	84.2	11 42	209 51		
21. —	83.3	66.9	78	81.1	12 28	210 54		
22. —	83.0	69.9	66	83.1	13 28	212 42		
23. —	82.9	71.1	45.0	83.7	13 52	214 49	Not observed.	
1. December	82.5	68.7	89	82.5	16 32	219 6		Between the Marianas and Philippines.
2. —	81.4	70.2	76.0	81.2	17 23	220 46		
3. —	81.5	71.8	61	79.6	18 25	222 54		
4. —	80.8	70.9	45	79.8	19 20	225 28		
5. —	79.0	67.1	73	79.8	19 44	227 45		
6. —	79.0	67.6	83	77.3	19 44	229 25		
11. —	82.0	60.0	95	82.0	18 0	240 7		In the Chinese Sea, on the west side of Luçon.
12. —	80.5	60.1	80.5	81.7	16 42	240 34		
13. —	82.2	61.5	93.5	84.5	13 51	240 24		
1818.								
24. March	71.9	62.7	145	76.1	33 14 S.	330 1		On the bank, and near the Cape of Good Hope.
26. —	71.1	64.0	54	72.0	34 2	331 48		
27. —	68.1	51.7	86	77.5	35 17	337 4		

Days.	Temperature of the Sea Water on the Surface.	Temperature of the Sea Water below the Surface.	Depth in Fathoms.	Temperature of the Air.	The Ship's Latitude. ° ′ ″	The Ship's Longitude. ° ′ ″	Transparency of the Water in Fathoms.	Observations.
1818.								
13. April	67.5	49.5	200	68.0	30 39 0 S.	345 33 W.	5	On the west side of the Cape of Good Hope
—	—	60.8	50	—	—	—	—	
—	—	66.1	25	—	—	—	—	
20. —	73.5	60.8	61.25	71.8	20 33 8	359 6 47	Not observed.	
22. —	73.	62.8	65.5	72.1	19 18 6	1 25 11		
23. —	73.7	58.1	54.5	75.0	17 55 22	3 8 15		
24. —	74.3	62.8	46	72.5	16 14 43	5 7 29		
26. —	75.6	62.	56.5	74.0	14 12 8	7 55 49		
27. —	77.2	59.8	61.33	78.3	12 30 52	9 58 4		
28. —	78.5	65.5	72	80.3	11 11 38	11 21 10		
29. —	79.4	60.1	70	80.7	9 39 38	12 46 6		
30. —	80.3	64.	61.25	80.3	8 15 54	14 3 0		
1. May	81.5	59.	56.5	81.7	6 35 32	15 34 0		
2. —	81.6	57.6	63	82.0	5 8 8	17 14 2		
3. —	82,6	56.	72.5	83.5	3 42 16	18 41 00		
4. —	83.	57.1	80	82.5	2 17 58	19 50 40		
5. —	83.	57.3	80	83.0	0 53 42	20 28 10		
6. —	83.3	59.1	56.5	84.1	0 7 58N.	20 26 17		
—	83.3	58	69.25	8.5	0 36 28	20 39 49		
7. —	—	—	—	—	—	—		
8. —	82.6	57.5	77.75	74.0	+1 58 7	+21 6 8		
9. —	84.3	58.5	80.	81.8	2 32 48	21 13 6		
10. —	84.5	58.9	80.	84.4	3 5.47	21 24 0		
11. —	83.	59.	77.25	79.3	3 30 11	+21 53 35		

Days.		Temperature of the Sea Water on the Surface.	Temperature of the Sea Water below the Surface.	Depth in Fathoms.	Temperature of the Air.	The Ship's Latitude.	The Ship's Longitude.	Transparency of the Water in Fathoms.	Observations.
1818.						° ′ ″	° ′ ″		
12.	May	82.5	59.2	80.	77.1	+ 4 16 7N	+22 42 48W	Not observed.	
13.	—	82.6	57.9	78.5	81.9	4 33 27	+24 11 23		
14.	—	82.9	56.5	79.75	83.9	5 29 46	26 9 34		
15.	—	81.9	55.9	69.	81.8	6 0 54	+27 34 28		
16.	—	81.5	58.	61.33	82.3	7 13 52	28 32 30		
17.	—	79.2	58.4	72.	80.0	9 27 17	+29 7 00		
18.	—	78.8	59.4	65.5	79.8	11 35 2	30 56 27		
19.	—	76.5	58.1	65.5	77.7	13 24 36	+32 2 41		
20.	—	76.5	65.5	64.	77.8	15 51 5	+32 56 8		
21.	—	78.8	68.6	72.	77.7	18 1 53	34 24 34		
22.	—	76.2	68.5	78.5	76.9	19 59 59	+35 10 10		
23.	—	75.8	68.8	61.33	76.7	21 40 20	36 14 49		
24.	—	76.9	69.6	78.5	77.0	23 6 7	36 51 0		
25.	—	76.	68.9	72.5	76.0	25 23 31	37 0 0		
26.	—	74.5	65.7	74.75	75.0	27 38 48	37 10 0		
27.	—	73.	66.5	61.33	75.5	30 3 49	37 24 0		
28.	—	72.	67.1	65.5	72.7	+32 36 58	+36 35 14		
29.	—	69.2	62.	75.25	69.3	34 34 31	35 55 36		
30.	—	74.5	62.3	74.25	69.9	35 41 48	+35 12 34		
31.	—	68.7	62.2	63.	73.0	37 9 11	34 31 34		
1.	June	68.9	61.5	74.25	68.7	38 9 11	33 8 12		
2.	—	67.5	60.1	72.	65.0	+39 15 43	31 3 0		
3.	—	66.7			63.0	+40 30 9	+29 47 50		
4.	—	64.1	58.6	73.75	64.0	+41 43 28	+27 23 20		

Days.	Temperature of the Sea Water on the Surface.	Temperature of the Sea Water below the Surface.	Depth in Fathoms.	Temperature of the Air.	The Ship's Latitude.	The Ship's Longitude.	Transparency of the Water in Fathoms.	Observations.
1818.					° ′ ″	° ′ ″		
5. June	63.5			64.0	+43 35 59 N	+24 9 56 W	Not observed.	
6. —	60.6	54.7	59.5	65.	45 57 1	21 23 00		
7. —	60.	54.5	67.	60.7	47 18 27	20 30 21		
8. —	60.3	54.7	77.	61.3	47 32 44	20 24 47		
9. —	62.5	54.2	75.25	60.4	48 1 43	19 42 49		
10. —	62.	54.2	78.75	62.3	48 1 32	17 56 57		
11. —	62.5	52.3	80.	63.5	48 8 10	15 33 6		
12. —	63.8	51.1	71.5	69.9	48 22 31	13 45 48		
13. —	59.5	52.1	82.	68.0	48 42 9	10 50 54		

REMARKS

ON THE PRECEDING OBSERVATIONS ON THE

SPECIFIC GRAVITY OF THE SEA WATER

IN DIFFERENT LATITUDES,

AND

ON THE TEMPERATURE OF THE OCEAN AT DIFFERENT DEPTHS.

BY J. C. HÖRNER.

THE observations on the specific gravity of the sea water have been already drawn up in an instructive table (page 403.) by the able naturalist of the expedition, and arranged according to the degrees of latitude. This table evidently shows the fact, which is also proved by the experiments on Krusenstern's voyage, that the sea on the surface, between the tropics, is specifically heavier, that it contains more salt, than in higher latitudes. If we take together the statements from 25th degree south, as far as 25th degree north latitude; and, in the same manner, from 50° to 65° degree of north latitude, the mean of the first is, 1.0288, that of the latter 1.0245, which gives the difference of 0.0043 or $\frac{1}{230}$. But this by no means proves an absolute inequality in the saltness of the water in general. To give a decided opinion on it, the sea water must be fetched up from considerable depths, and weighed. Probably the greater saltness arises from the rapid decrease of the fresh water, in consequence of evaporation. From the well known slowness of the transition of chemical elements in undisturbed compounds, this decrease is but slowly repaired; and as the upper layers are also the warmer, they may, notwithstanding their greater specific density, in consequence of their extent, be maintained by the warmth swimming above the lower cooler layers, by which a principal agent of commixture, the difference of weight, is rendered of no effect. This slowness of change, and the condensation of the saline solution at the surface, which results from it, has the advantage, that the acceleration of the evaporation sets bounds to itself, because, with the increasing condensation, the attraction of the salt to the parts of the water is greater, and, consequently the diminution of the latter less. Without this arrangement, the tropical seas would perhaps be covered like the frozen seas of the north, with constant fogs. Subsequent experiments will show how far our explanation of this inequality is correct; of which we have now more hopes, as convenient accurate apparatus have been discovered to fetch up water from any depth, at pleasure, and unmixed.

The considerable number of observations (there are one hundred and sixteen of them) on the temperature of the sea below the surface, their extent over waters of the ocean remote from each other, and probably, also, their accuracy, give them a decided claim to the attention of the natural philosopher; and the perseverance with which they were continued, under various circumstances, does honour as well to the Naturalist of the Expedition, as to the commander, who not only in calms, but in some periods, almost daily, afforded the necessary assistance. They were all made with the Six-thermometer *, which is a good assurance of their accuracy. It is certainly remarkable, that an instrument so simple, so convenient in the use, so certain in the results, and which has been long known, is not more frequently used for this purpose; so that in the latest scientific voyages, much more uncertain thermometers have been used, to which only the deep sea clamm of Captain Ross forms an exception.

Our observations fall under two heads: measurements of the temperature in different depths, in the same places of the ocean, and in statements of the warmth in the usual soundings, from sixty to eighty fathoms, in different places.

The most complete observations on the changes

* This instrument was divided, after the English manner, in degrees of Fahrenheit: I have reduced the result to degrees of Reaumur.

of the temperature, in increasing depths, are, in the South Sea, of the 13th and 14th of September, 1817, in 36° north latitude, and 148° west longitude. Besides confirming the general law, that the cold increases with the depth, they also afford the following results.

1. The upper parts of the water show a particular warmth, as the temperature, in the first eight fathoms, diminished only 0°, 4 R., but from that depth to twenty-five fathoms, full 6° R. From twenty-five fathoms to a hundred fathoms' depth, the decrease of warmth is considerably less, since, in the next twenty-five fathoms, it is only 1°, 7 R., and in the next fifty fathoms, only 1°, 5 R.; a decrease, which amounts to only the tenth part of the preceding. It is still slower between a hundred and three hundred fathoms.

2. If we compare these observations with those of the 6th of June, 1816, in 37° north, and in 199° west longitude, consequently, in the same parallel of latitude, the influence of the season is particularly observable in the temperature, on the surface, which in June is 13° R., in September 18° R. It, however, does not go much deeper than from twenty-five to fifty fathoms; and at a hundred fathoms it is already within the limits of the accuracy of such observations; for we have,

For 100 fathoms	6th June,	9°, 4 R.
	13th September,	9, 4 —
	14th September,	8, 6 —

3. A certain coincidence with these results, only on a greater scale, is shown by the experiments of the 15th of Nov. 1817, in 9° north latitude, and 205° west longitude, in which the temperature decreases from the surface to about sixty or seventy fathoms, rapidly and uniformly, from 24°,7 R. to 8°, 8 R. From 69 to 101 fathoms, this rapid decrease, instead of proceeding, is suddenly reduced to the small amount of 0°, 9 R. But if we compare these observations with those immediately preceding and succeeding them, of the 13th, 14th, and 17th of November, we shall hesitate to draw from them decisive conclusions.

The observations of 13th April, 1816, in 15° south, and 130° west, follow a quite different course from those in September, 1817, in 36° north. The decrease of warmth from the surface, to as far as a hundred fathoms' depth, is much more inconsiderable, being here only 3°, 6, there nearly treble, namely, 9°, 4 Reaum. It becomes more considerable between a hundred and two hundred fathoms, namely, 8°, 8 R. Remarkable as this inequality is, it yet seems impossible to ascribe it to an error in the observation, such as too soon drawing up the thermometer; for, on the one hand, the regular course of the experiments of the 14th September, 1817, and their coincidence with those of the 13th, at the depths of 0, 25, and 100 fathoms, does not allow us to suppose any thing of the kind; on the other side, the observations of

13th April, 1816, find their confirmation in the preceding ones of the 7th of April, in 18° south, which give a difference of 0 to 125 fathoms of 4°, 8 R., that is, from 0 to 100 fathoms; likewise 3°, 8 R. The same observations then give for the second hundred of the depth in fathoms, likewise about 8° Reaumur.

It is not to be discovered, from the observations, whence this difference in the progressive decrease of the warmth arises. It cannot well be ascribed to the influence of the seasons, at least in latitude 35° north: the observations of June and September, show an agreement with each other. The reason perhaps is, that the perpendicular rays of the sun penetrate the water, between the tropics, to a greater depth than in latitudes where the sun never appears in the zenith. The place of constant temperature, independent of the seasons, must, probably, lie much deeper between the tropics than beyond them.

5. The observations of the 22d of September, 1817, in 28° north latitude, and in 152° west longitude, seem to present a much more uniform course, particularly if we set aside the statement in twenty-five fathoms' depth, which does not appear to agree with the higher or lower observations. We have from them a decrease of heat, of 3°, 5 R. for the first fifty fathoms; 3°, 0 R. for the second fifty fathoms, and 4° 3 R., from a hundred to two hundred.

The collective observations on the progress of the decrease of heat were made in the South Sea. From the Atlantic Ocean we received only a few insulated statements, for depths of a hundred to two hundred fathoms. The experiments in both oceans are arranged in the following table.

WARMTH of the SEA WATER at different Depths, arranged according to the Geographical Latitudes in degrees of Reaumur's Thermometer.

Month.	Surface.	70 to 90 Fathoms.	100 Fathoms.	200 Fathoms.	300 Fathoms.	Latitude.	Longitude.	
April	21.0	—	17.2	9.9	—	18 S.	125 W.	
—	21.4	—	17.8	10.8	—	15	134	
May	22.6	—	13.5	—	—	1 N.	177	
Nov.	24.5	—	10.7	—	—	9	204	In the South Sea.
—	23.0	14°.1	—	—	—	12	210	
Dec.	22.1	12.8	—	—	—	16	240	
—	21.7	16.6	—	—	—	18	224	
Sept.	20.1	—	13.0	8.8	—	28	152	
June	18.7	—	13.5	—	9.4	29	199	
Sept.	18.0	—	9.3	7.0	5.4	36	147	
					(4.8)*			* In 400 Fathoms.
June	13.0	—	9.3	—	5.0	37	199	
Jan.	10.4	—	—	3.2	—	44 S.	57	
March	17.3	—	12.3	—	—	34	27	In the Atlantic Ocean.
April	15.8	12.8	—	7.9	—	31	15	
Oct.	18.9	—	10.6	—	—	30 N.	15	
—	16.4	—	—	11.0	—	39	13	

The temperatures in the usual soundings from 70 to 80 fathoms, appear, on account of their considerable number, from which mean numbers may be deduced, the best calculated to supply fundamental data. Yet some singular results appear in them. Among these is the statement in the South Sea, that in 18° N. latitude, and 76 fathoms' depth, in December, the water was $2\frac{1}{2}$° R. warmer than in 11° N. latitude, and 70 fathoms' depth, in November. Perhaps the local places of observation have had here some influence. The observation in 11° latitude lies in the west of the Mariana islands, and in the north of the Philippines, consequently sheltered against the warmer currents from the south, by a kind of wall, and open only to the north, while on the other hand, the place, in 18° latitude, lies more in the open sea. The temperature, found at a depth of 90 fathoms, in the Chinese Sea, to the west of Luçon, is remarkably cold; perhaps in consequence of the north-east currents prevailing in December.

Almost daily observations on the temperature were made in the Atlantic Ocean, from the 20th of April to the 13th of June, 1818, mostly at a depth of 70 fathoms. In order to balance the possible errors of the observations which may arise from the difference in the time that the thermometer was under water, I have added several together, and noted the mean number. They are in the following table. The figures in parentheses show the number of observations the mean of which is given.

Observation.		Temperature of the Water on the Surface.	Temperature of the Water below the Surface.	Depth in Fathoms.	Latitude.	Longitude.
					° ′	° ′
April 20–26.	(5)	18.6	13.0	57	17 15 S.	3 20 W.
— 27–30.	(4)	20.8	13.5	66	10 24	12 2
— 30–May 4.	(5)	22.1	11.8	67	5 12	17 5
May 3–10.	(8)	22.7	11.4	74	0 43 N.	20 28
— 10–16.	(7)	22.6	11.4	75	4 51	24 38
— 15–1 9.	(5)	21.2	11.5	67	9 34	29 38
— 20–24.	(5)	20.3	16.1	71	19 30	35 7
— 25–30.	(6)	18.3	14.8	71	31 0	36 30
— 31–June 6.	(5)	15.1	12.3	68	40 30	29 40
June 7–13.	(7)	13.2	9.6	77	48 9	17 15

This table shows a similar anomaly to that which we noticed in the South Sea. That is, the proportionately low temperature near the equator, from 5° S. to 10° N. Perhaps the greater heat between 20° and 30° of southern latitude, might be a remnant of the southern summer. But the considerable increase of temperature in the zone, between 15° and 30° north latitude, is still more remarkable. For though, towards the end of May, the sun was near the zenith of those parts, yet this influence, which could be only commencing here, must have shown itself in the waters near the equator, which the sun had just traversed at the time of those observations (in April), which was by no means the case. The temperatures at the surface indicate indeed this influence of the sun, being the highest at the equator ($22\frac{1}{2}$° R.), while the southern half of the tropical seas had already assumed an autumnal temperature, since we observe here, in 17° S., the same warmth (18½° R.) as in 30° N.

The difference of the seasons, of the places, and of the depths in which these observations were made, allow us neither an accurate comparison of them among each other, nor with the small number of facts collected by other observers so as to enable us to deduce general and definite rules. Desirable as it would be that these examinations of the temperature of the air should be made in certain parallels, for example, from 10° to 10° degrees, circumstances cannot always be commanded, though the excellent series of daily observations which the preceding table contains, show what may be done without waiting for calms. But the observers might at least agree to make all their experiments at the same depths, namely, at 25, 50, 100, 200 fathoms, in which the frequent deviation of the plumb-line from the perpendicular direction must be taken into account. Frequent repetitions of the observations, at every opportunity, is the more advisable, because, in unexpected and anomalous results, it is the only means of guaranteeing their accuracy; and we would particularly recommend attention to a sufficient duration of the experiment, and to frequent comparison of the thermometrograph with an accurate thermometer, at different temperatures. Lastly, the more detailed experiments on the progressive decrease of heat at small intervals, if they are made in distant latitudes, different temperatures, and seasons, are a valuable acquisition to the doctrine of heat in general, and in particular to the natural philosophy of our globe.

POSTSCRIPT.

THE author of the "Remarks and Opinions," being at a distance from the commander and narrator of the expedition, was unable to compare and correct his statements or decisions by those of his learned coadjutors. He could not even correct his orthography of foreign names and words to make it agree with that observed in the journal of the voyage, as he did not see the proof-sheets of the work. With respect to written languages, he has followed the authority of native authors; and with respect to the unwritten ones, principles of his own, of which he has given an account in his note to the vocabulary.

Many of these papers, in the interval between their origin and their publication, in the progress of time and knowledge, are already superseded by later information. The author wished to have cancelled them. South America has been brought nearer to us. Important works and constant intercourse have opened Brazil to us. Chili is no longer the country we saw; we present a picture of the past; free trade now exports the copper which the first defenders of independence were obliged to use for cannon balls.

Later discoveries have brought the question which we discussed respecting the polar regions nearer to a decision, and advanced the point of view from which it is to be contemplated. Lieutenant Parry has penetrated from Lancaster Sound between islands and masses of land intersected by channels, as far as the 115° of west longitude (an extent of 35°,) only 20° this side of the meridian of Mackenzie River. We are inclined to imagine that similar islands and masses of land occupy a great part of the polar regions between Greenland and New Siberia, and particularly in the north of Beering's Straits, (Burney.)

On the other hand, the discovery of New South Shetland, by William Smith, in 1819, which we cannot help fancying is united with Sandwich land, has revived the belief of a southern continent, to which Cook himself was attached, even after his second voyage. This coast, bordering on one of the most frequented seas, and several hundreds of ships, on account of the westerly storms on the west passage, are obliged to approach it within a few degrees; we are astonished at the lateness of the discovery.

Lastly, W. Scoresby (An Account of the Arctic Regions. Edinburgh, 1820,) has given us a work on the North Polar Regions, the solid reasoning of which throws our hasty sketch into the shade.

These essays appear unchanged. And the author, removed from the place of publication, is not able to remedy the defects of which he is sensible. He will only add a few corrections and remarks.

D. ADELBERT VON CHAMISSO.

In March, 1821.

CORRECTIONS AND REMARKS.

VIEW OF THE GREAT OCEAN, &c.

Tagalese Literature.

F. C. Alter, on the Tagalese Language, Vienna, 1802, merely informs us that an imperfect MS. Tagalese Vocabulary exists in the Imperial Library at Vienna.

"System of Languages and Numerals of the Eastern Islands of the Great Ocean." When we wrote our observations on the dialects of the islanders of the Great Ocean, we had not been able accurately to compare the dialect of Tonga with any other dialect of the same general language, and such a comparison was necessary sufficiently to establish our decision.

We must here return our thanks to a man of learning, who took a lively interest in the object of our research, and zealously strove to procure for us the literary assistance which we required. His Excellency the Minister of State, Baron William Von Humboldt exerted himself to procure for us some of the books which the venerable missionaries in the Society islands have written in the same language, and which have been printed partly at Paramatta, in New South Wales, partly in Otaheite itself, and of which mention is made in the Narrative of the Mission at Otaheite. London, 1818.

We see, with astonishment, these islands, under the influence of Christianity, quickly and tranquilly rising from a state of social order, resembling ours in the middle ages, to that which has but now begun for our world after long and bloody storms. The people and the chiefs there take each other by the hand over the ruins of the abolished social system, of the taboo, and of arbitrary power; the written law is unanimously and solemnly desired, proposed, and confirmed, and the foreign teachers, who refrain from all interference in the concerns of the state, behold, with prayers of gratitude, the growth of the seed they have sown.

While we hoped, in vain, for specimens of the rising literature of Otaheite, our wish has been fulfilled in another dialect, and we are indebted for it to the same beneficent Missionary Society. We have before us a Grammar and Vocabulary of the Language of New Zealand, published by the Church Missionary Society. London, 1820. 8vo. The author of this grammar is the same Mr. Kendall who has communicated to us the Vocabulary in Nicolas's voyage. The language has now been opened to us, and we correct our opinion.

The dialect of New Zealand has, like the Tonga, pronouns of the three persons in the singular, and the four persons in the dual and plural, (we mean the double first person, of which the one includes, and the other excludes, the person addressed.) The pronouns of the dual are formed of the roots of those of the plurals, and of the number two. All appear, in the dialect of New Zealand, more simple and more concentrated than in the dialect of Tonga, where every person has several pronouns of different uses. These pronouns, and particularly those of the twofold first person of the plural, must be the most difficult part of the language for a foreigner, what he last conceives and makes himself master of. Being an essential part of the Malay language, they may, perhaps, exist in all the dialects of Eastern Polynesia, and we now believe we have done wrong in omitting, as dubious, in the dialect of Owhyee, the pronoun of the third person, which Lisianskoy mentions. It is *Oyera*, which coincides with *Iy-a*, Malay; *Siya*, Tagalog; *Ia*, Tonga and New Zealand.

The particles which mark the time and mode of the action, are different in the dialects of Tonga, New Zealand, and Owhyee.

It is very far from easy to find out the arithmetical system of a people. It is at New Zealand, as at Tonga, the decimal system. What may, perhaps, have deceived Mr. Kendall, at the beginning, in his first attempt in Nicolas's voyage, and which we followed, is the custom of the New Zealanders to count things by pairs. The natives of Tonga count the bananas and fish likewise by pairs and by twenties (*Tecow*, English score). The decimal and vigesimal system frequently run into one another, (*quatre-vingt, six-vingt, quinze-vingt.*) We do not believe we have erred, with respect to Radack, but the numeral system of the Owhyeeans, and other people of the Great Ocean, may, perhaps, require a more accurate examination.

The orthography fixed in the New Zealand Grammar, is natural and worthy of recommendation; and it is to be hoped that it agrees with that followed in the Otaheite books.

MANILLA. — VOLCANO DE TAAL.

THE drawing mentioned by us of the crater of the Volcano de Taal, will be found in the "Voyage Pitoresque," which M. Choris is going to publish at Paris, under the especial patronage of his Highness Count Romanzoff. This beautiful and faithful gallery of our voyage will greatly illustrate our observations and remarks. We often found it superfluous to describe what it was the province of the able artist to represent to the eye.

THE END.

LONDON:
Printed by A. & R. Spottiswoode,
New-Street-Square.

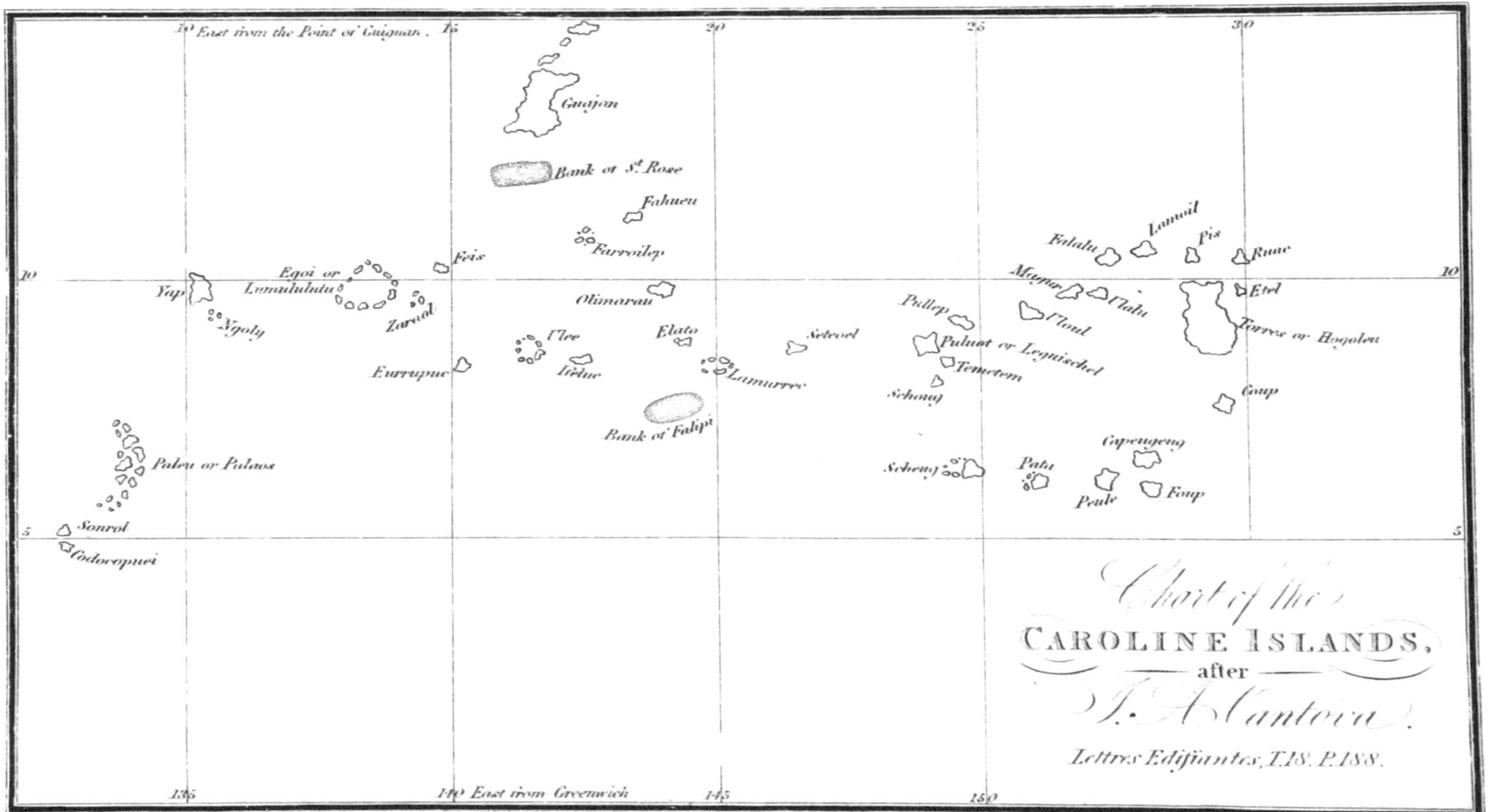

London, Published by Longman & Co. Octr. 8th 1821.
Sidy. Hall, Sculpt.

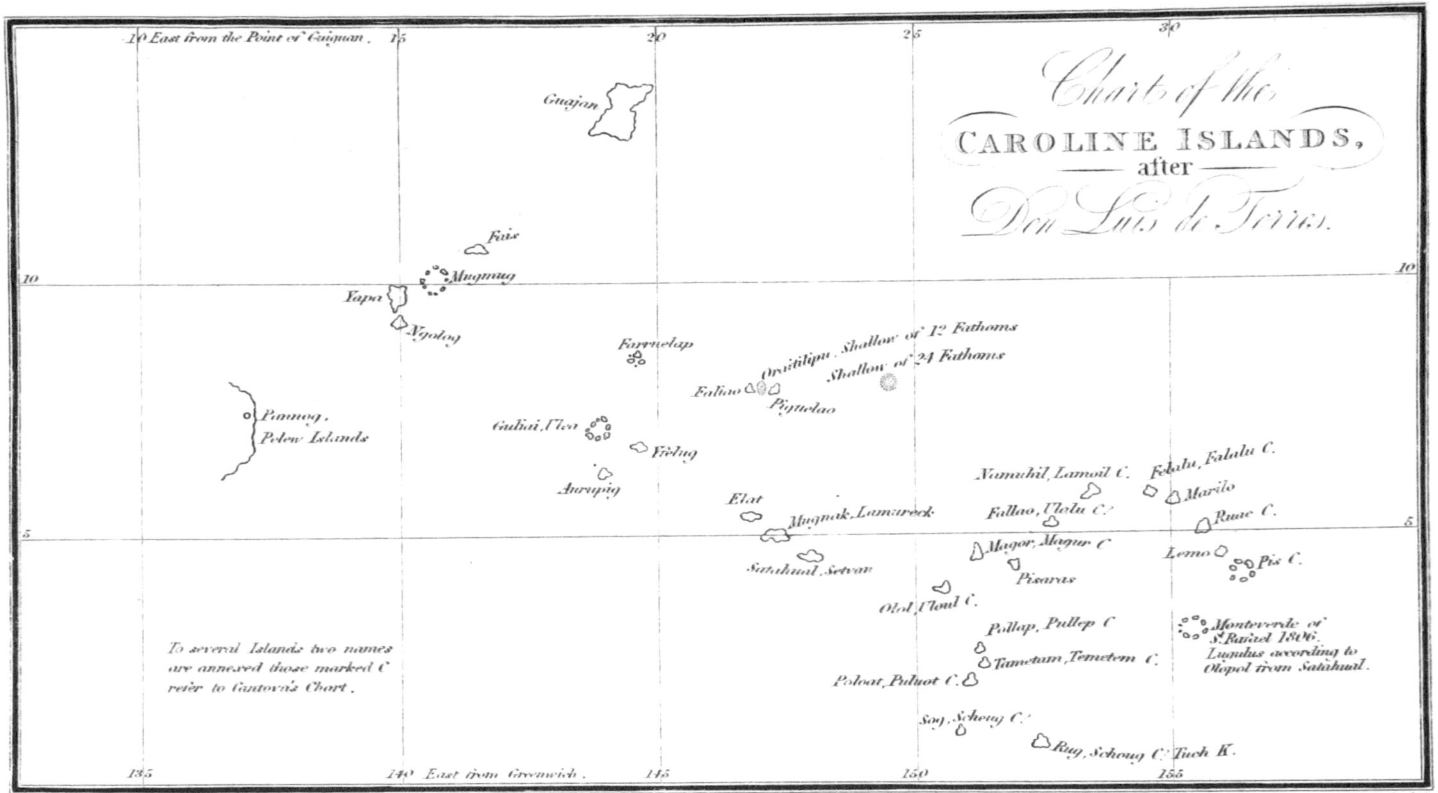

Sid. Hall, Sculp.

London Published by Longman & Co. Oct. 8th 1821.

For EU product safety concerns, contact us at Calle de José Abascal, 56–1°, 28003 Madrid, Spain or eugpsr@cambridge.org.

www.ingramcontent.com/pod-product-compliance
Ingram Content Group UK Ltd.
Pitfield, Milton Keynes, MK11 3LW, UK
UKHW040226290926
15471UKWH00001B/12
* 9 7 8 1 1 0 8 0 5 7 5 9 2 *

Cambridge University Press has long been a pioneer in the reissuing of out-of-print titles from its own backlist, producing digital reprints of books that are still sought after by scholars and students but could not be reprinted economically using traditional technology. The Cambridge Library Collection extends this activity to a wider range of books which are still of importance to researchers and professionals, either for the source material they contain, or as landmarks in the history of their academic discipline.

Drawing from the world-renowned collections in the Cambridge University Library and other partner libraries, and guided by the advice of experts in each subject area, Cambridge University Press is using state-of-the-art scanning machines in its own Printing House to capture the content of each book selected for inclusion. The files are processed to give a consistently clear, crisp image, and the books finished to the high quality standard for which the Press is recognised around the world. The latest print-on-demand technology ensures that the books will remain available indefinitely, and that orders for single or multiple copies can quickly be supplied.

The Cambridge Library Collection brings back to life books of enduring scholarly value (including out-of-copyright works originally issued by other publishers) across a wide range of disciplines in the humanities and social sciences and in science and technology.

CAMBRIDGE LIBRARY COLLECTION

Books of enduring scholarly value

Maritime Exploration

This series includes accounts, by eye-witnesses and contemporaries, of voyages by Europeans to the Americas, Asia, Australasia and the Pacific during the colonial period. Driven by the military and commercial interests of powers including Britain, France and the Netherlands, particularly the East India Companies, these expeditions brought back a wealth of information on climate, natural resources, topography, and distant civilisations. Their detailed observations provide fascinating historical data for climatologists, ecologists and anthropologists, and the accounts of the mariners' experiences on their long and dangerous voyages are full of human interest.

A Voyage of Discovery, into the South Sea and Beering's Straits, for the Purpose of Exploring a North-East Passage

An admirer of Captain Cook, Otto von Kotzebue (1787–1846) was a leading navigator, in Russian service, circumnavigating the globe three times. His 1815 expedition set out to find a passage through the Arctic, study the coastlines of Kamchatka and Alaska, and explore the Pacific. Among the personnel were the naturalist Chamisso and the artist Choris, who both contributed valuable information to the published account, while Eschscholtz, a physician, collected zoological specimens. Originally published in 1821 in Russian and German, this English translation, presented with many plates and charts, appeared the same year and formed part of Darwin's library aboard the *Beagle*. Volume 3 continues with Chamisso's detailed observations as well as contributions by other members of the expedition. Flora and fauna are identified, mineral samples taken, and aerometric readings recorded along with the temperature of the sea.